International Conference on Environmental, Industrial and Energy Engineering (EI2E 2017)

IOP Conference Series: Earth and Environmental Science
Volume 104

Shanghai, China
19 – 22 October 2017

ISBN: 978-1-5108-5493-2
ISSN: 1755-1307

Printed from e-media with permission by:

Curran Associates, Inc.
57 Morehouse Lane
Red Hook, NY 12571

Printed by Curran Associates, Inc. (2018)

For permission requests, please contact the Institute of Physics
at the address below.

Institute of Physics
Dirac House, Temple Back
Bristol BS1 6BE UK

Phone: 44 1 17 929 7481
Fax: 44 1 17 920 0979

techtracking@iop.org

Additional copies of this publication are available from:

Curran Associates, Inc.
57 Morehouse Lane
Red Hook, NY 12571 USA
Phone: 845-758-0400
Fax: 845-758-2633
Email: curran@proceedings.com
Web: www.proceedings.com

Table of contents

Volume 104

2017 International Conference on Environmental, Industrial and Energy Engineering (EI2E 2017)

19–22 October 2017, Shanghai, China

Accepted papers received: 18 December 2017
Published online: 30 December 2017

IOP Conf. Series: Earth and Environmental Science **104** (2017) 011001 doi:10.1088/1755-1315/104/1/011001

PREFACE

EI2E 2017 was held in Shanghai, China during October 19-22, 2017. EI2E 2017 was organized by Asia Pacific Institute of Science and Engineering (APISE). The conference provides a useful and wide platform both for display the latest research and for exchange of research results and thoughts in broad Environmental, Industrial and Energy Engineering. The participants of the conference were from almost every part of the world, with background of either academia or industry, even well-known enterprise. The success and prosperity of the conference is reflected high level of the papers received.

The proceedings are a compilation of the accepted papers and represent an interesting outcome of the conference. This book covers 2 chapters: Energy Engineering; Environmental and Industrial Engineering.

We would like to acknowledge all of those who supported EI2E 2017. Each individual and institutional help were very important for the success of this conference. Especially we would like to thank the organizing committee for their valuable advices in the organization and helpful peer review of the papers.

We sincerely hope that EI2E 2017 will be a forum for excellent discussions that will put forward new ideas and promote collaborative researches. We are sure that the proceedings will serve as an important research source of references and the knowledge, which will lead to not only scientific and engineering progress but also other new products and processes.

Dr. Yaser Qudaih, Kyushu Institute of Technology, JAPAN
Prof. Nopbhorn Leeprechanon, Thammasat University, Thailand

IOP Conf. Series: Earth and Environmental Science **104** (2017) 011002 doi:10.1088/1755-1315/104/1/011002

COMMITTEES

Conference Chairs

Dr. Yaser Qudaih, Kyushu Institute of Technology, JAPAN

Prof. Nopbhorn Leeprechanon, Thammasat University, Thailand

Program Co-Chair

Prof. Yongsheng CHEN, The Chinese University of Hong Kong, HK

International Technical Committees

Assoc Prof. Dr Aeslina Abdul Kadir, Universiti Tun Hussein Onn Malaysia, Malaysia

Dr. Grzegorz Lisak, Nanyang Technological University, Singapore

Dr. Sayed M. Bateni, University of Hawaii at Manoa, USA

Dr. Huichao Chen,Southeast University, China

Dr. Wenxian Yang, Newcastle University, UK

Dr. SITI NAZAHIYAH BTE RAHMAT, Universiti Tun Hussein Onn Malaysia, Malaysia

Dr. Ivan CK Tam, Newcastle University, UK

Dr. NORZILA BTE OTHMAN, Universiti Tun Hussein Onn Malaysia, Malaysia

Dr. Hong Yue, University of Strathclyde, UK

Dr. Manousos Valyrakis, University of Glasgow, United Kingdom

Dr. HARTINI BINTI KASMIN, Universiti Tun Hussein Onn Malaysia, Malaysia

Dr. Zhibin Yu, University of Glasgow, United Kingdom

Dr. Goh Hong Ching, Universiti Malaya, Malaysia

Dr. Chunhua LIU, The City University of Hong Kong, Hong Kong

Dr. Hassan Migdadi, American university of Madaba, Jordan

Peer review statement

All papers published in this volume of *IOP Conference Series: Earth and Environmental Science* have been peer reviewed through processes administered by the proceedings Editors. Reviews were conducted by expert referees to the professional and scientific standards expected of a proceedings journal published by IOP Publishing.

The Combustion Performance and Ingredient Ratio of Thermite

Shuan-zhu Jia[1,*], Shi-guo Du[1], Jian-wei Zhen[1] and Xin-hui Yang[2]

[1] School of Ordnance Engineering College, Shijiazhuang 050003, China
[2] Troops of 31679, China

*E-mail: 810566134@qq.com

Abstract. Thermite was widely used because of its combustion properties and combustion products. However, due to the combustion heat, ignition performance, burning rate and the ratio of energy conversion of different components of thermite agent are very different. The requirements of the main realization in: (a) Its easy to ignite and not easy to extinguish; (b) Combustion and heat as much as possible High; (c) The burning speed should be appropriate. So the performance of these aspects is always being hot focus. In this paper, four aspects of the improve about combustion heat, ignition performance, burning rate and the ratio of energy conversion were analyzed through the aluminum alloy, the addition of aluminum, the addition of the third party, the change of the particle size and the compaction density. Finaly states the research direction in the future. The future of aluminum heat agent formula research focus will be: (a) A single aluminum heat agent the best proportion of the study; (b) The addition of different additives and additives (c) The exploration of alternatives that are more excellent performance will inevitably become a hot topic to improve the heat of combustion. Aluminum heat agent performance will be much superior, and the application will also be more extensive.

1. Introduction

Thermite is widely used because of its excellent performance of high temperature and the ability to produce molten metal. The initial range of thermite is narrow, indicating only Al/Fe_2O_3 applied at that time, however, with the expansion of the scope of application, resulting in a number of different application conditions and special needs. So as to improve the combustion performance of the new requirements, therefore need to change the aluminum heat agent formula to achieve the aluminum heat modified. All along, the aluminum heat agent components of the inquiry have never stopped. The aluminum heat agent combustion performance has become more and more in-depth.

2. Heat of Combustion

2.1 Single Aluminum Heat Agent

For different aluminum heat, the reaction heat is not the same, the following is the standard conditions of different oxide powder and aluminum powder reaction occurs when the reaction:

$$2Al + Fe_2O_3 \rightarrow 2Fe + Al_2O_3 \quad + 856.6kJ/mol \tag{1}$$
$$2Al + Cr_2O_3 \rightarrow 2Cr + Al_2O_3 \quad + 535.6kJ/mol \tag{2}$$
$$2Al + 3CuO \rightarrow 3Cu + Al_2O_3 \quad + 1212.5kJ/mol \tag{3}$$

The above reaction exotherm is the standard molar reaction calorific value, which reflects the energetic size of each aluminum heat agent, for the choice of aluminum heat to provide a basic basis.

However, for the aluminum heat agent, only the high heat of reaction can not explain the high reaction temperature, Wang Yi and so on according to the principle of itching balance ratio, and through the use of sol/gel method prepared nano-Al/Fe_2O_3 Aluminum heat agent, measured the heat release of 1648J/g. Tillotson T M measured the total reaction heat of nano-Al/Fe_2O_3 at 1500J/g [1], and Wang Yi's conclusion is basically consistent. The molar value of the molar reaction is about 672 kJ/mol. Can be seen between the real heat and the ideal heat is still a big gap between. So simply press

According to the chemical reaction equation for a simple ratio does not necessarily achieve the best combustion effect. The difference in the composition of the main components will lead to different ignition temperature, burning speed and other factors, which lead to different macro temperature. So the single aluminum heat agent in the proportion of the study will still be an indispensable future important direction.

2.2 Mixed with a Variety of Aluminum Heat Agent

A single aluminum heat agent has a specific number of combustion properties, ignition performance, etc., when trying to mix different aluminum heat agent, the expected combustion performance can be a variety of aluminum heat agent characteristics of the integrated, to meet the needs of different occasions. CuO can produce a lot of oxygen during the reaction, and the reaction heat of Al/CuO is higher than that of iron-aluminum heat, and the metal jet produced by the reaction is more stable, but the unsuitable content is too high because the generated high pressure is prone to explosion, So the amount of hot cut can be appropriate reference. Al/Fe_2O_3 reaction speed is slow, and combustion stability, so suitable for the need for stable combustion in the occasion. Zhang Jing and so on by adjusting the cutting agent aluminum heat agent formula after the test concluded: Al/Fe_2O_3 burning point is too high, which can not produce the spray slag; Al/CuO burning speed is too fast, the pressure is too large; Al/Y (which is easy to ignite and generate high temperatures while reducing gas generation due to the absence of gas) can increase combustion heat and make combustion more stable; Al/Y increases its flammability due to its low ignition and the release of large amounts of heat Sex. When the above formula is controlled at 30%: 30%: 20%: 20% can achieve the best cutting effect. Similar experimental results, Wang Peng and so on through the experimental study of pyrotechnic cutting agent of the best melt metal aluminum heat agent formula: Al/Fe_2O_3 can reduce the combustion propagation rate, Al / CuO can produce high pressure oxygen to generate Metal copper jet, Al/X can produce a lot of heat to increase the system reaction temperature. When the content of Al / Fe_2O_3 is 30% ~ 50%, Al/CuO is 30% ~ 50%, then 10 ~ 20% Al/X is added, the melting effect is the best. Wang Sen [2] and so on through the inorganic thermodynamic analysis and calculation of the cutting agent aluminum heat agent formula to explore the choice of Al/Fe_2O_3 and Al/CuO mixture, Al/CuO in the mass fraction of 13.6% to 22.8% When the best cutting effect, Al/CuO is too high working time is too short, Al/CuO is too low temperature is not enough.

Shen Lianhua et al. determined the combustion properties of Al/B/Fe_2O_3 nanocomposite energetic materials. The results show that when Al content accounts for 80% of the reducing agent, B accounts for 20% of the reducing agent, that is, Al accounts for 44.3% of the total Al/B/Fe_2O_3 nanocomposites, 44.7% of Fe_2O_3, and 11.1% of the total mass. , With a high reaction enthalpy.

It can be seen from the above, in the development of aluminum heat agent, the study of the combustion performance of different aluminum heat and the effect of mixed aluminum heat on the overall performance of the study more and more in-depth, aluminum heat agent performance will be more suitable for different areas The demand.

2.3 Change the Particle Size and Join the Third Party Material

In recent years, with the improvement of processing technology, nano-level metal particles have been greatly improved, and around the nano-level aluminum heat agent performance research also will be carried out, through the "super aluminum heat" is also known Scholars on the nano-aluminum heat agent affirmed.

Sullivan K [3] and so on by ultrasonic dispersion method in the Al/CuO aluminum heat agent added nano-B to test its performance and combustion performance, found that nano-B molar content of less than 50% can enhance its combustion performance. An Ting [4] with hydrothermal preparation of MnO_2 nanotubes and nano-aluminum powder as raw materials prepared nano-aluminum heat agent, combustion heat up to 6165kJ/ kg. Since the hydrogen content in $NaBH_4$ is 10.66% [5], $NaBH_4$ has a significant effect on the improvement of combustion energy. In the formulation of the combustion machine, $NaBH_4$ is added to $NaBH_4$ by orthogonally The results showed that the addition of $NaBH_4$ could significantly increase the calorific value, and when the content of $NaBH_4$ was 20%, the increase was up to 14.3%. It can be seen that changing the particle size and the addition of additives to the heat of the aluminum heat of the heat can be seen in the future will inevitably become a hot way to improve the heat of combustion.

3. Ignition Performance

Aluminum ignition agent ignition temperature is high, not easy to ignite, resulting in the practical application of ignition is not very convenient, which to some extent limit the aluminum heat agent as a promotion of combustion, so reduce the ignition temperature for the application of aluminum heat Extraordinary meaning.

3.1 Single Aluminum Heat Agent

For a single aluminum heat agent, by changing the content of some ingredients in the formula, to achieve the purpose of improving the ignition temperature is a more direct and effective method. For the aluminum heat agent in a major component - the effect of oxidant changes on the combustion performance, is the focus of discussion of scholars. Xue Yan prepared Al/MoO_3, and its combustion heat is 8351J/g, and the impact sensitivity is low, the flame sensitivity is more sensitive. Zheng Baoming [6] and so on according to the reaction product integrated DSC curve analysis $KMnO_4$ in aluminum thermal reaction in the actual reaction is $KMnO_4$ decomposition products MnO_2. And Li Qanqan [7] and so on with frozen ball milling method prepared Al/MnO_2 composite powder, and that it is more than other methods of preparation of the degree of mechanization activity is higher, more reactive, the reaction heat release is also significantly increased.

For the effect of aluminum powder on the ignition performance, Umbrajker [8] and so on by measuring the ignition temperature of Al/MoO_3 system found that at the same heating rate, the ignition temperature with the Al content increases, the ignition temperature will be increased. Similarly, Bazyn [9] et al through the shock wave ignition experimental study found that: $Al-MoO_3$ ignition delay period with the increase in Al content and longer. It can be seen that the impact of aluminum powder for the same positive correlation for the aluminum heat agent ratio provides a basis.

With the development of scientific research, the understanding of various raw materials continues to deepen. At the same time due to the complex process of nano-aluminum powder preparation, so there has been a substitute for Al powder exploration, France [10] carried out on the nano-P/CuO aluminum heat agent research, found that the aluminum heat agent impact sensitivity moderate, Sensitivity and electrostatic fire sensitivity is high, similar to conventional detonating drugs, and red phosphorus is easy to obtain, and more easily oxidized, and CuO reactivity stronger, so it is expected in the future a lot of applications. The future for more excellent performance of the alternatives will be an important development direction.

3.2 Change the Particle Size and Join the Third Party Material

Bazyn [11] et al through the shock wave ignition experiments found that: nano-Al powder delay period for the microsecond; and micron-level Al powder delay period is longer, in milliseconds. Indicating that the use of nano-level of raw materials easier to ignite. This indicates the direction of the ignition modification of the nano-scale aluminum heat agent.

It is possible to reduce the ignition temperature of the aluminum heat agent by adding some substances which are relatively easy to ignite in the aluminum heat agent and which can produce a higher temperature to ignite the aluminum heat agent. According to the Al-Mg/$KMnO_4$ aluminum heat, It was found that when the content of Mg was higher, the lower the ignition temperature and lower from 723K to 493K. Zhou Jianzhong [12] and so on in the aluminum heat agent by adding Si-Bi_2O_3 mixture can increase the heat of aluminum heat agent while reducing the ignition temperature, and when the content of 20% when the minimum ignition temperature of 613 ℃. France [13] developed a kind of aluminum in the aluminum alloy by adding a special structure of graphitized carbon black can reduce the sensitivity of aluminum heat agent. All of which illustrate the third-party material for the ignition temperature reduction has a very important role.

3.3 Density

The pressing density is also one of the factors affecting the ignition delay period. Stamatis [14] prepared by the ARM method 8Al-MoO_3 system of nano-aluminum heat agent pressed into a high-density cylinder, the laser beam as the ignition source to test its reactivity. The results show that the ignition delay period becomes longer with the pressing density. The higher the pressing density leads to the higher the thermal conductivity, the greater the specific heat, resulting in longer ignition time.

4. Speed of Burning

4.1 Single Aluminum Heat Agent

Al powder as a reducing agent for aluminum thermal reaction, its impact on the combustion rate has always been taken into account. Prakash [15] the effect of Al/$KMnO_4$ on the combustion performance was determined by the difference in the mass fraction of Al/KMnO4. When the mass fraction of Al was 35%, the reaction rate reached the highest. Dutro GM [16] prepared nano-Al/MoO_3 aluminum heat agent, and by changing the composition of these different components into the combustion tube and found: When the mass of Al from 10% to 65%, the aluminum heat agent combustion reaction Speed from 100m/s stability to 1000m/s; greater than 85% or less than 10% are not a good reaction rate. It can be seen that Al powder has a significant effect on the burning rate of aluminum heat.

4.2 Adding Third Party Material

Third-party material can change the microstructure of aluminum heat agent composition, as the intermediate reaction medium and produce gas pressure to speed up the reaction rate and so on. Yang [17] et al. prepared nano-Al powder coated MnO_2/SnO_2. Experimental results show that the reaction temperature is reduced, because SnO_2 can reduce the surface free energy of MnO_2, make nano-Al and contact MnO_2 more closely, Increase the delivery speed. Wang Zhi and so on by adding NaF in Al/Fe_2O_3, found that with the increase of NaF, the rate of self-propagation decreased first and then increased, the early part of the absorption of heat, play a role in reducing the reaction, the latter is F The role of ions on the surface of Al on the destruction of Al_2O_3 play a role in the destruction of hormones. Wang Sen [18] and so on through the test to determine the impact of KNO_3 on thermally cutting ammunition, in the isovolumetric environment, due to the presence of KNO_3 gas pressure, which will speed up the burning rate, but the resulting slag spray will have a favorable impact on cutting. The experimental results show that when KNO3 is 4.8% ~ 7.8%, the cutting effect is the best.

4.3 Change the Pparticle Size

The contact area between the nanomaterials is large, so the conduction rate of the reaction increases. This is also true for aluminum heaters. Sun J [19] et al. studied the thermal behavior of Al/MoO_3 prepared with different sizes of Al powder particles. The results show that the activity of nano-Al particles is higher than that of microns, and with the increase of nano-Al particle size The reaction rate increases. Weismiller MR [20] et al. prepared Al/CuO and Al/MoO_3 nano-aluminum heat by the method of physical mixing. The experimental results show that the combination of nano-Al powder and nano-oxide is the fastest, Comparative experiments show that nano-oxide nano-Al powder than the contribution of the contribution rate of large. In the test of Al/B/Fe_2O_3 combustion performance, it was found that the nanocomposite $Al/B/Fe_2O_3$ was 6 times higher than the physical blends of $Al/B/Fe_2O_3$, and the sensitivities were obviously more than physical blends $Al/B/Fe_2O_3$ low.

5. Rate of Energy Conversion

For a substance, the size of its utilization determines the level of efficiency, for enterprises, the savings and reduction of raw materials related to the vital interests. Therefore, the size of the energy conversion rate for the efficient use of aluminum heat and the actual application of the amount of loading has a crucial role. Umbrajkar [21] et al. measured the atmospheric pressure change by constant ignition test to estimate the energy conversion efficiency in the Al/MoO_3 combustion process. The results show that when the ratio of raw materials is 4 and 8, the energy conversion efficiency is much higher than that of aluminum powder. When the ratio of raw materials is 12 and 16, the energy conversion efficiency nanometer Al powder is equivalent, and the reason is aluminum High, the greater the particle size of the complex, leading to this phenomenon. Schoenitz [22] et al through experimental studies found that Al/MoO_3 in the raw material ratio of 8, the fastest energy release rate. Stamatis [23] et al. demonstrated that the energy conversion effect was best when the 8Al-3CuO complex with a mass fraction of 20% was added to the pure aluminum powder. Thus, Al powder on the aluminum heat of the energy conversion rate also has a significant impact.

6. Discussion and Prospect

The main combustion properties of aluminum heat agent include combustion exotherm, ignition performance, burning rate, energy conversion rate, etc. The requirements of the main realization in: (a) Its easy to ignite and not easy to extinguish; (b) Combustion and heat as much as possible High; (c) The burning speed should be appropriate. The improvement of the above performance is mainly through the size of the raw material, oxidants, reducing agents and additives selection and the ratio, compaction density and other aspects of the change to achieve, so the study of aluminum heat agent on the need for a large number of Experiment to support.

With the understanding of the types and properties of materials continue to deepen, the future of aluminum heat agent formula research focus will be: (a) A single aluminum heat agent the best proportion of the study; (b) The addition of different additives and additives (c) The exploration of alternatives that are more excellent performance will inevitably become a hot topic to improve the heat of combustion. Aluminum heat agent performance will be more and more superior, the application will also be more extensive.

References

[1] Tillotson T M, Gash A E, Simpson R L, et al. Nano structured energeic materials using sol-gel methodlogies[J]. 2001 *J Non-Cryst Solids* **285 (1-3)** 338.

[2] Wang Sen, Xin Wentong, Wu Yongsheng, Shi Lifeng. Effect of Thermit on Cutting Capability of Combustion Cutting Ammunition [J]. 2012 *Casting·Forging·Welding* **41 (13)** 202-204. (in Chinese)

[3] Sullivan K, Young G, Zachariah M R. Enhanced reactivity of nano-B/Al/CuO MIC 's [J]. 2009 *Combustion and Flame* **156** 302-309.

[4] An Ting, Zhao Fengqi, Ma Haixia, Ren Xiaoning, Zhao Ningning, Wang Qiong, Yang Yong. MnO_2 Nanotube and Its Super Thermite: Preparation and Their Effect on Thermal Decomposition of Cyclotrimethylene Trinitramine[J]. 2015 *CHINESE JOURNAL OF INORGANIC CHEMISTRY* **32 (1)** 97-102. (in Chinese)

[5] Gui Dayong, Liu Jiping, Dai Lan. Applicationof Sodium Borohydride ($NaBH_4$) in IncendiaryAgent [J]. 2007 *CHINESEJOURNALOFENERGETICMATERIALS* **15 (3)** 231-234. (in Chinese)

[6] Zheng Baoming, Jiang Wei, Chen Wei, Li Fusheng. Study on Performance of Super-reactive $AlMg/KMnO_4$ Thermites [J]. 2013 *ACTA A RMAMENTA RII* **34 (9)** 1161-1166. (in Chinese)

[7] Li Qanqan, Chen Yueguang, Wang Shijiao, Fan Runhua, Guo Jingyan, Yao Disheng. Synthesis

and Characterization of Al/MnO_2 Composite Powder through Cryomilling [J]. 2013 *JOURNAL OF SYNTHETIC CRYSTALS* **42 (5)** 950- 952. (in Chinese)

[8] Umbrajkar S M, Schoenitz M, Dreizin E L. Control of structural refinement and composition in Al-MoO_3 nanocomposites prepared for misassedness [J]. 2006 *Propellants, Explosives, Pyrotechnics* **31 (5)** 382 -389.

[9] Bazyn T, Lynch P, Krier H, et al. Combustion measurements of fuel-rich aluminum and molybdenum oxide nano-composite mixtures [J]. 2010 *Propellants, Explosives, Pyrotechnics* **35 (2)** 93-99.

[10] Comet M, Siegert B, Schnell F, et al. Phosphous-based nanothermites: a new generation of pyrotechnics illustrated by the example of n-CuO / Propellants Explos. 2010 *Pyrotech* **35**.

[11] Bazyn T, Glumac N, Krier H, et al. Reflected shock ignition and combustion of aluminum and nanocomposite thermite powder [J]. 2007 *Combustion Science and Technology* **179 (3)** 457 -476.

[12] Zhou Jianzhong, Tian Jin, Xue Guo. Influence of Si-Bi2O3 on the Ignition Performance of Al-CuO Thermite [J]. 2016 *www.gdchem.com Guangdong Chemical Industry* **43 (12)** 58-59. (in Chinese)

[13] Arnaud Bach, Pierre Gibot. Modulation of the reactivity of a WO_3/Al energetic material with graphitized carbon black as additive. 2015 *46th Int. Annual Conference of ICT*.

[14] Stamatis D, Dreizin E L. Thermal initiation of consolidated nanocomposite thermites [J]. 2011 *Combustion and Flame* 2008.

[15] Prakash A, Mccormick A V, Zachariah M R. Synthesis and reactivity of a superreactive matastable intermolecular composite formulation of Al / KMnO4 [J]. 2005 *Adv Mater* **17 (7)** 900.

[16] Dutro G M, Yetter R A, Risha G A, et al. The effect of stoichiometry on the combustion behavior of a nanoscale Al/MoO_3 thermite [J]. 2009 *Proceedings of the Combustion Institute* **32** 1921-1928.

[17] Yang Y, Zhang Z-C, Wang P-P, et al. Hierarchical MnO_2/SnO_2 heterosturctures for a novel free-standing ternary thermite membrane [J]. 2013 *Inorganic Chemistry* **52 (16)** 9449-9455.

[18] Wang Sen, Xin Wen-tong, Wu Yong-sheng, Shi Li-feng, Wu Yi-ying. Study on Influence of KNO_3 on Cutting Capability of Combustion Cutting Ammunition [J]. 2012 *INITIATORS & PYROTECHNICS* **(2)** 14-17. (in Chinese)

[19] Sun J, Pantoya M L, Simon S L. Dependence of size and size distribution on reabtivity of alminum nanoparticles in economic with oxygen and MoO3 [J]. 2006 *Thermochimica Acta* **444 (2)** 117-127.

[20] Weismiller Mr, Malchi J Y, Lee J G, et al. Effectsof fuel and oxidizer particle dimedsions on the propagation of aluminum containing thermites [J]. 2011 *Proceeddings of the Combustion Institute* **33 (2)** 1989-1996.

[21] Umbrajkar S M, Seshadri S, Schoenitz M, et al. Aluminum-rich Al-MoO3 nanocomposite manufactured prepared by columns reactivemilling [J]. 2008 *Journal of Propulsion and Power* **24 (2)** 192-198.

[22] Schoenitz M,Ward T S,Dreizin E L. Fully dense nano-composite energetic powders prepared by arrested reactive milling[J]. 2005 *Proceedings of the Combustion Institute* **30(2)** 2071 — 2078.

[23] Stamatis D,Jiang X,Beloni E,et al. Aluminum burn rate modifiers based on reactive nanocomposite powders [J]. 2010 *Propellants,Explosives,Pyrotechnics* 3**5 (3)** 260 -267.

Effect of Structural Parameters on the Combustion Performance of Platelet Engines

Yin Liang[1,*] **and Liu Weiqiang**[1]

[1]Science and Technology on Scramjet Laboratory, National University of Defense Technology, Changsha, Hunan 410073, People's Republic of China

*Corresponding e-mail: yl88222@126.com

Abstract: Numerical simulation was adopted to determine its flow and combustion characteristics by using gaseous methane and oxygen as the main propellants, the effects of nozzle space and expanding angle are examined for the single element splash platelet injector. Navier-Stokes (*N-S*) equations were solved for the gas-gas flow field with a reduced mechanism involving 9 species and 1 reaction. Results indicated that large corner recirculation zones are produced in the combustor head. This phenomenon consequently enhances mixing and stabilizes combustion, but non-uniformity in temperature contour is observed in the combustor head. Recirculation zone decreases as nozzle space increases, which induces the decrease of maximum temperature and high temperature regions, but it has little influence on the combustion efficiency and combustion length. The combustion length and maximum temperature decrease initially and then increase as expanding angle increases. Conversely, a D value of 2.4 mm and γ value of 60° are selected for the future works because of the shortest combustion length and minimum temperature of the injector faceplate.

1. Introduction

Injector design plays a critical role in the design of a liquid rocket engine. Designed by Aerojet in the 1970s, platelet injector is a new type of liquid rocket injector, which are widely used in low thrust attitude control engine because of their unique features, including excellent response, simple structure, potential high performance with reduced chamber length [1, 2]. The face of a platelet injector is formed from a stack of thin metal platelets which are diffusion bonded into a single unit. The platelet concept was selected because of several features including ease of manufacture and infinite feature geometries with precise tolerance control [3-5], which can provide solutions to difficult thermal and fluid flow problems [6]. Over the past decades, considerable efforts have been dedicated in order to get the baseline performance statistics of platelet injector [7-9].

Boyd [10-13] had an introductory description of the Orbit Maneuvering Engine (OME), followed by a description of the approach chosen for OME uprating, including uprating ground rules and issues. Details values of the current and uprated OME are presented in table 1. The results shown that higher chamber pressure and expansion ratio have significant influences on specific impulse. The injector design used the concept of a platelet injector, the face is formed of twelve photo etched platelets which are bonded together into a 0.088 inch thick stack which is then bonded and welded to the body. Of the 450 elements, 378 are "splash-plate"" elements, with 72 "X-doublet" elements in the outer row. Fuel film cooling is provided by 36 "splash-plate" orifices along the outer row.

Table 1 A comparison of the current and uprated OME

	Current OME	Updated OME
Thrust, lbf	6000	6000
Chamber pressure, psia	125	350
Expansion ratio	55:1	154:1
Mixture ratio	1.65	1.95
Specific impulse, s	315	334
Dry weight, lbm	297	322

Aerojet Company undertook to evaluate the AJ10-221[13-14] for Orbit Maneuvering Engine in 1985. This engine employs a platelet injector design that includes 92 pre-atomized splash platelet elements. The injector is fabricated from CRES 347 using a photochemical machining and diffusion bonding process. A high-temperature indium/rhenium chamber was also used in these new-generation engines. During the research on AJ10-221, several key technologies are applied:

(1) A high-temperature indium/rhenium chamber was selected because it provides good low-temperature ductility and high melting point.

(2) Splash platelet elements were employs for the injector pre-atomized design. The pre-atomized element avoids the problem of requiring perfect alignment/impingement of two small diameter liquid streams for the purpose of propellant atomization.

(3) The companion technologies for providing metallurgical joints between the different materials used for the injector, chamber, and high area ratio skirt.

This paper presents a numerical analysis of a splash platelet injector using gaseous oxygen and methane as propellants on the ANSYS Fluent platform. In this work, effects of nozzle space (*D*) and expanding angle (γ) on single element injector are studied. A comparison of combustion flow fields and component distribution are given. This study revealed the baseline performance statistics of a single element injector, which can provide a reference for the structural optimization of the platelet engines.

2. Physical method

The schematic of the splash platelet element and its fixed parameters are shown in Figure 1. The key parameters that affect the flow and combustion characteristics include oxidizer orifice (D_O), fuel orifice (D_f), nozzle space (D), expanding angle (γ), length of trapezoidal orifice (L_l), thickness of the middle plate ($H2$), and thickness of the faceplate ($H3$). The design parameters of the chamber thrust are listed in Table 2.

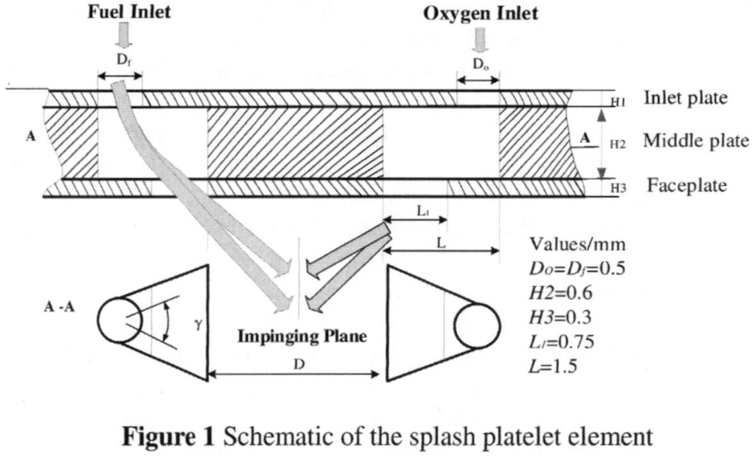

Figure 1 Schematic of the splash platelet element

Table 2 Design parameters of the chamber thrust

Parameters	Values
Chamber pressure/MPa	0.85
Mixture ratio(O/F)	3.2
Mass flow rate *m*/(g/s)	44
Gas temperture/K	298.15
Chamber diameter /mm	24.28
Throat diameter /mm	10.86
Theoretical characteristic velocity /(m/s)	1827.3
Ground Specific impulse /(m/s)	2270.2

3. Numerical methods

3.1 Governing equations

The three-dimensional, Navier-Stokes and species transport equations for a chemically reacting gas of *N* species are given by [15]:

$$\frac{\partial(E-E_V)}{\partial x}+\frac{\partial(F-F_V)}{\partial y}+\frac{\partial(G-G_V)}{\partial z}=H \tag{1}$$

Where *E, F, G* are convective flux vectors

$$E=[\rho u, \rho uu+p, \rho uv, \rho uw, u(\rho e+p), \rho uY_i]^T \tag{2}$$

$$F=[\rho v, \rho vu, \rho vv+p, \rho vw, v(\rho e+p), \rho vY_i]^T \tag{3}$$

$$G=[\rho w, \rho wu, \rho wv, \rho ww+p, w(\rho e+p), \rho wY_i]^T \tag{4}$$

$$H=[0,0,0,0,0,w_i]^T \tag{5}$$

$$E_v=[0,\tau_{xx},\tau_{xy},\tau_{xz},u\tau_{xx}+v\tau_{xy}+w\tau_{xz}-q_x,\rho_i D_{im}\,\partial Y_i/\partial x]^T \tag{6}$$

$$F_v=[0,\tau_{yx},\tau_{yy},\tau_{yz},u\tau_{yx}+v\tau_{yy}+w\tau_{yz}-q_y,\rho_i D_{im}\,\partial Y_i/\partial y]^T \tag{7}$$

$$G_v=[0,\tau_{zx},\tau_{zy},\tau_{zz},u\tau_{zx}+v\tau_{zy}+w\tau_{zz}-q_z,\rho_i D_{im}\,\partial Y_i/\partial z]^T \tag{8}$$

$$\tau_{xx}=[-\frac{2}{3}\mu(\frac{\partial u}{\partial x}+\frac{\partial v}{\partial y}+\frac{\partial w}{\partial z})+2\mu\frac{\partial u}{\partial x}] \qquad \tau_{yy}=[-\frac{2}{3}\mu(\frac{\partial u}{\partial x}+\frac{\partial v}{\partial y}+\frac{\partial w}{\partial z})+2\mu\frac{\partial v}{\partial y}]$$

$$\tau_{zz}=[-\frac{2}{3}\mu(\frac{\partial u}{\partial x}+\frac{\partial v}{\partial y}+\frac{\partial w}{\partial z})+2\mu\frac{\partial w}{\partial z}] \qquad \tau_{xy}=\tau_{yx}=\mu(\frac{\partial u}{\partial y}+\frac{\partial v}{\partial x})$$

$$\tau_{yz}=\tau_{zy}=\mu(\frac{\partial w}{\partial y}+\frac{\partial v}{\partial z}) \qquad \tau_{xz}=\tau_{zx}=\mu(\frac{\partial w}{\partial x}+\frac{\partial u}{\partial z})$$

According to the Fick's law, the following equations can be obtained:

$$\rho_i u_i=\rho D_i\frac{\partial Y_i}{\partial x} \tag{9}$$

$$\rho_i v_i=\rho D_i\frac{\partial Y_i}{\partial y} \tag{10}$$

$$\rho_i w_i=\rho D_i\frac{\partial Y_i}{\partial z} \tag{11}$$

3.2 Turbulence Model

A compressible standard $\kappa-\varepsilon$ is used for the turbulence model. The turbulence kinetic energy, κ, and it rate of dissipation, ε, can be obtained from the following transport equations:

$$\frac{\partial(\rho k)}{\partial t}+\frac{\partial(\rho k u_i)}{\partial x_i}=\frac{\partial}{\partial x_j}\left[\left(\mu+\frac{\mu_t}{\sigma_k}\right)\frac{\partial k}{\partial x_j}\right]+G_k-\rho\varepsilon \tag{12}$$

$$\frac{\partial(\rho\varepsilon)}{\partial t}+\frac{\partial(\rho\varepsilon u_i)}{\partial x_i}=\frac{\partial}{\partial x_j}\left[\left(\mu+\frac{\mu_t}{\sigma_\varepsilon}\right)\frac{\partial\varepsilon}{\partial x_j}\right]+C_{1\varepsilon}\frac{\varepsilon}{k}G_k-C_{2\varepsilon}\rho\frac{\varepsilon^2}{k} \tag{13}$$

The turbulent viscosity, μ_t, can be computed by combining κ and ε as follows:

$$\mu_t=\rho C_\mu\frac{\kappa^2}{\varepsilon} \tag{14}$$

Where, $C_{1\varepsilon}=1.44$, $C_{2\varepsilon}=1.44$, $C_\mu=0.09$, $\sigma_\kappa=1.0$, $\sigma_\varepsilon=1.3$

3.3 Boundary condition

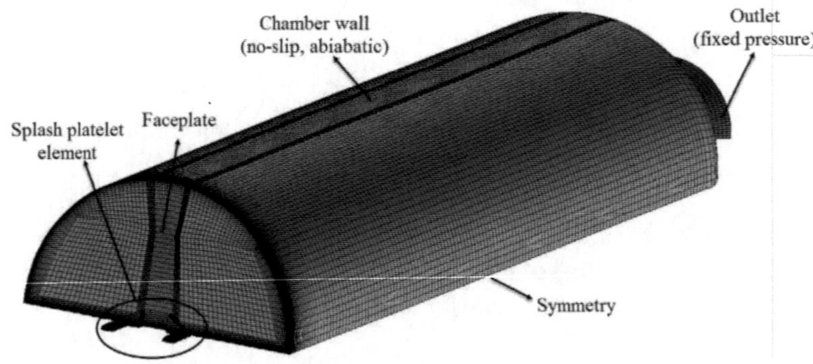

Figure 2 Grid of the calculation domain

Figure 2 shows the grid of the calculation domain and boundary condition. A symmetry boundary condition was adopted for the single element injector. For the boundary conditions, the inlets of methane and oxygen are characterized by a fixed mass flow rate, and the inlet turbulence intensity is set at 5%. The inlet temperatures of methane and oxygen are set at 298.15 K. The nozzle exit is specified as a supersonic outlet, and adiabatic nonslip wall boundaries are enforced at chamber walls [16, 17]. In the solution method, the coupling scheme with second order upwind discretization for all transported scalars is used.

3.4 Chemical kinetics mechanism

Navier-Stokes equations and the standard $\kappa-\varepsilon$ turbulence model are used to calculate the combustion flow field of the combustor. In the present simulations, a detailed O_2/CH_4 chemical kinetics mechanism involves dozens of chemical species with hundreds of elementary reactions. A reduced mechanism, which contains 9 species and 1 reaction, is utilized in this study to save computing time. The interaction between turbulence and reaction was considered by the EDC model [18, 19].

$$CH_4+1.52O_2\rightarrow 0.33CO_2+0.662CO+1.454H_2O+0.37H_2+0.134H+0.216OH+0.047O \tag{15}$$

4. Result and Discussion

4.1 Validation of numerical methods

Table 3. Design parameters of the shear-coaxial injector

Propellant	*m/(g/s)*	*MR*	*P_C/MPa*	*D_C/mm*	*D_t/mm*	L^* */mm*	R_v
GCH_4/GO_2	5.5	3.5	3	26	14.8	1180	1.77

To demonstrate the validity of numerical methods and chemical kinetics model, simulation was performed on a shear-coaxial GO_2/GCH_4 injector, where P_C is the chamber pressure, MR is mixture ratio, D_c is chamber diameter, D_t is throat diameter, L^* is the characteristic length, R_V is the velocity ratio. The parameters of the shear-coaxial injector are listed in Table 3, and the detail design values can get from Ref.[20]. Figure 3 shows the temperature distribution of the oxidizer post, which is consistent with the Ref. data; it can exactly describe the temperature distribution although there is little difference. This testified the accuracy of numerical methods and chemical kinetics model. Detailed chemical reaction model should be adopted in the future works.

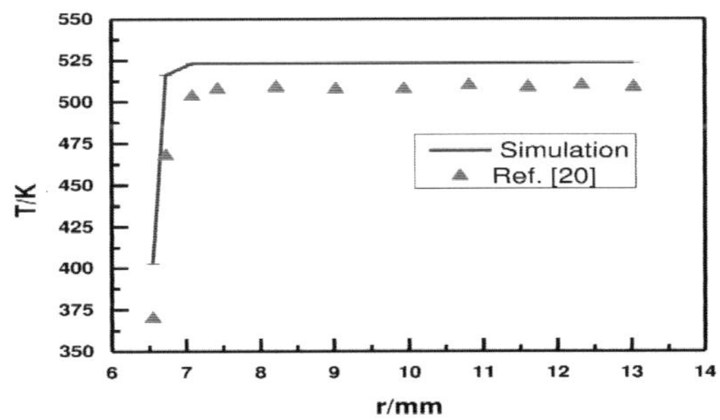

Figure 3. Temperature distribution of the oxidizer post

4.2 Effect of D on the combustion characteristics

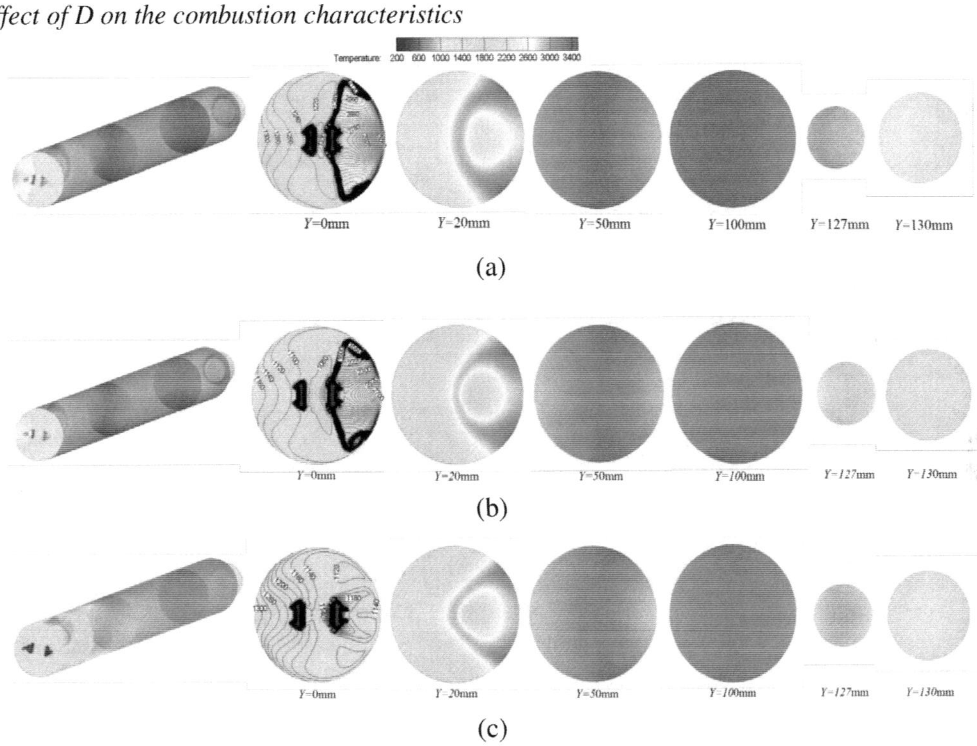

Figure 4 Temperature distributions for different nozzle spaces: (a) D=1.2mm; (b) D=1.8mm; and (c) D=2.4mm

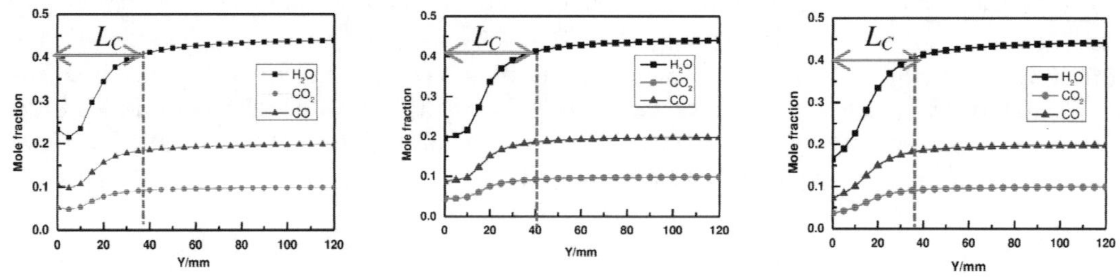

(a) D = 1.2 mm (b) D = 1.8 mm (c) D = 2.4 mm

Figure 5 Effect of nozzle spaces on H_2O, CO_2 and CO mole fraction

When studying the effect of nozzle space D, expanding angle γ is fixed as 60 degree. Figure 4 shows the average temperature distribution in the cross section. As calculated in the simulation, the combustor temperature is as high as 3400 K, which is similar to the theoretical value. D has significantly influences on the temperature in the chamber head. As D increase, the recirculation zones on the front part of the combustor and gas temperature decrease. These recirculation zones appear mainly at the position of Y= [0, 30 mm] away from the injector faceplate, which induces mixing enhancement and combustion stability. The influence of recirculation zones decrease with the development of the combustion process, which leads to more well-distributed temperature in the combustor.

For methane combustion, the proportion of water is larger than that of other components. Therefore, the main combustion area and combustion length can be revealed by analysing the mole fraction of water. In this study, the combustion length is defined as L_C when the mole friction of water is 90% of the theoretical value.

The effect of D on CO_2, CO, H_2O mole fraction is shown in Figure 5. As D increase, the mole fractions increase rapidly and then reach a balance in a shorter combustion length. The CO_2, CO, H_2O mole fraction make slightly differ with D=1.2, 1.8, and 2.4mm. When D= 2.4mm, the corresponding combustion length is about 40mm.

4.3 Effect of γ on the combustion characteristics

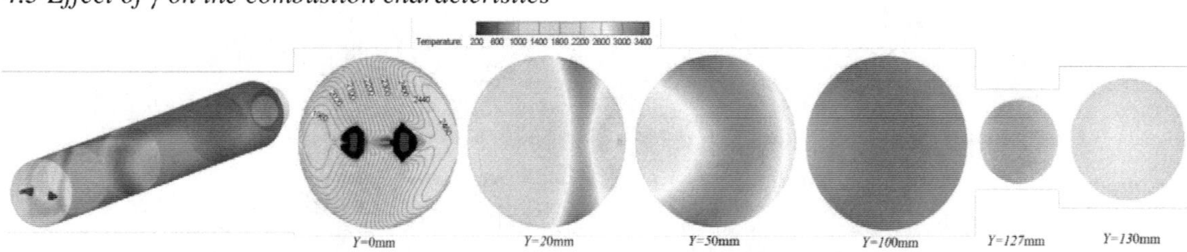

(a)

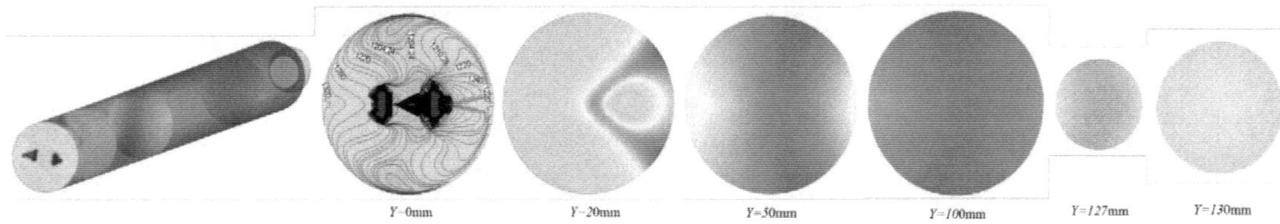

(b)

IOP Conf. Series: Earth and Environmental Science **104** (2017) 012002 doi:10.1088/1755-1315/104/1/012002

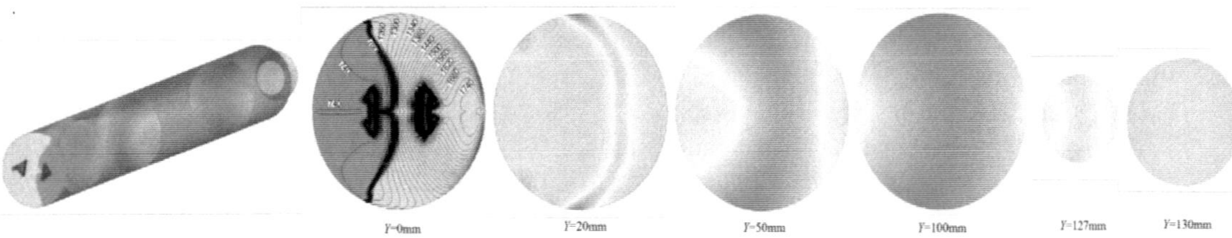

(c)

Figure 6 Temperature distributions for different expanding angles: (a) $\gamma = 30°$; (b) $\gamma = 45°$; and (c) $\gamma = 75°$

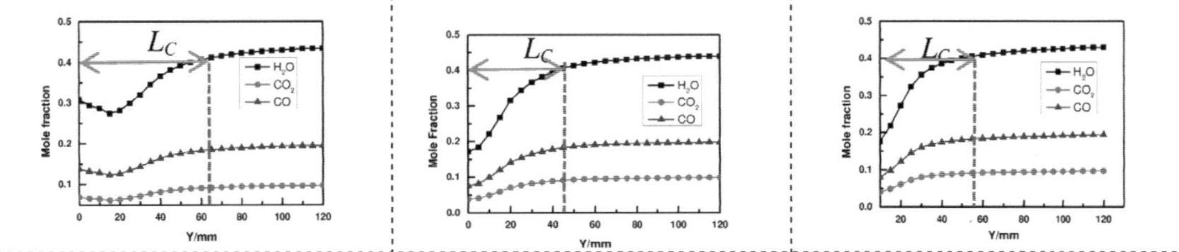

(a) $\gamma = 30°$ (b) $\gamma = 45°$ (c) $\gamma = 75°$

Figure 7 Effect of expanding angles on H_2O , CO_2 and CO mole fraction

When studying the effect of expanding angle γ, nozzle space *D* is fixed as 2.4mm. In the follow-up study, γ is selected as a decision variable, where γ=30, 45, and 75 degree. As γ increases, the flow area increases, but the velocity ratio remains the same. Figure 6 shows the temperature distributions for different expanding angle. Non-uniform Temperature is found in the combustion chamber head because of the characteristics of the splash platelet element. Compared with the effect of *D,* expanding angle has greater influence on the temperature distributions in the combustor head.

The effect of γ on the CO_2, CO, H_2O mole fraction is shown in Figure 7. Results show that the combustion length initially decreases and then increase as γ increase. The mole fraction of H_2O is quite different in the combustion chamber head, the combustion length are 76mm, 55mm, 62mm, respectively. A comparison with the above results shows that when γ= 60° , the combustion length is the shortest.

4.4 Temperature of the injector faceplate

Aerodynamic heating remarkably affects the reliability of injector faceplates. The effect of the aerodynamic heating of platelet injector can be more severe than those of traditional injector because of several unique features, such as shorter chamber length and thinner faceplate. A large-corner recirculation zone produced in the combustion chamber head because of the characteristics of the splash platelet element, which induces mixing enhancement and combustion stability but creates a heat flux peak on the injector faceplate. Figure 4 and 6 (Y= 0mm) shows the temperature distributions of injector faceplate. The results show that the high temperature region and the maximum temperature decrease as *D* increase. When D=1.2mm, the maximum temperature can be 3000 K, which exceeds the limit value of 1Cr18Ni9Ti (T<1400K). The maximum temperature initially decreases and then increases as γ increase. When γ = 45°, the maximum temperature is not exceeding 1400K. In all of the simulations, the temperatures of the injector faceplates exceed their limit. Therefore, faceplates with cooling techniques should be designed.

5. Conclusions

The combustion characteristics of a single element splash platelet injector with GO_2/GCH_4 as propellant were discussed through CFD simulation, and the effects of D and γ were examined. The following conclusions can be obtained:

(1) As D increase, the recirculation zones on the front part of the combustor and gas temperature decrease. Nozzle space has significant influences on the flow field in the combustor head, but it has little influences on the composition distribution and combustion length.

(2) As γ increase, the flow area increase, but the velocity ratio remains the same. Non-uniform temperature is found in the combustion chamber head, which has significant influence on the combustion length, the combustion length initially decreases and then increase as γ increase.

(3) The high temperature region and the maximum temperature decrease as D increase, and the maximum temperature initially decreases and then increases as γ increase. In all of the simulations, the temperatures of the injector faceplates nearly exceed their limit. Therefore, faceplates with cooling techniques should be designed.

Acknowledgment
This work is supported by National Science Foundation of China (Grant NO. 90916018) and Hunan Provincial Natural Science Foundation (Grant NO. 13JJ2002)

References

[1] H. H Mueggengurg. Platelet injector design and development history. AIAA/SAE/ASME/ASEE 27th Joint Propulsion Cong, 1991.

[2] KahL R C, Labotz R J, Bassham L B. Platelet Injector for Space Shuttle Orbital Maneuvering Engine, AIAA/SAE 10th Propulsion Cong, 1974.

[3] Liu W Q, Sun J, Nie T, et al. A structure of platelet heat-pipe-cooled leading edge, China Patent, ZL201310311505.5[P]. 2013-12-04.

[4] S.A. Poniaev, R.O. Kurakin, A.I. Sedov, et al. Hypervelocity impact of mm-size plastic projectile on thin aluminum plate Original Research Article, *Acta Astronautica*, 135(2017):26-33.

[5] Liu H P, Liu W Q. A numerical model for the platelet heat-pipe-cooled leading edge of hypersonic vehicle, *Acta Astronautica*, 118(2016) : 210-217.

[6] B. A. Robbers, B. J. Anderson, W. A. Hayes, et al. Platelet devices-limited only by one’s imagination, 42nd AIAA/ASME/SAE/ASEE Joint Propusion Conference & Exhibit 9-12 July 2006, Sacramento, California.

[7] Donald M J, Sanders D R, Leonard S. Durability Testing of the AJ10-221 490N High performance (321 Sec Isp) Engine, AIAA/SAE/ASME/ASEE 29th Joint Propulsion Conference and Exhibit, June 28-30, 1993, Monterey, CA.

[8] Sanders D R, Leonard S. New generation of high-performance engines for spacecraft propulsion, *Journal of Propulsion and Power*, Vol.10, No. 1, Jan.-Feb. 1944.

[9] D. David. Space Shuttle Maneuvering Subsystem (OMS) Rocket Engine Development Status Update July 1977, AIAA/SAE 13th Propulsion Conference, July 11-13, 1977, Orlando, Florida.

[10] Boyd, W, and Mallini C. Shuttle Performance Enhancement Using An Uprated OMS Engine, AIAA/ASME/SAE/ASEE Joint Propulsion Conference Boston, Massachusetts, July, 1988

[11] J. D. Bertolino, W. C. Boyd. Uprated OMS Engine Status-Sea Level Testing Results, N91-24256.

[12] Boyd, W., and Brasher, W. Uprated OMS Engine Status and Future Applications, AIAA/ASME/SAE/ASEE 22nd Joint Propulsion Conference, Huntsville, Alabama, June, 1986.

[13] Rosenberg, S. D., and Schoenman, L. New Generation of High Performance Engines for Spacecraft Propulsion, *Journal of Propulsion and Power*, 10(1): 40-46, 1994.

[14] Schoenman, L, Rosenberg, S. D., and Jassowski, D. M. Test Experience, 490-N High performance [321-s Specific Impulse] Engine, *Journal of Propulsion and Power*, 11(5): 992-997, 1995.

[15] Jin, P, Xu X, and Cai G B, Numerical simulation of combustion chamber of FFSC with gas-gas injectors [C], IAC-05-C4.P.18, 56*th* International Astronautical Congress, Oct 17-21, 2005, Fukuoka, Japan.
[16] Wang X W, Cai G B, Jin P. Scaling of the flowfield in a combustion chamber with a gas–gas injector, *Chin. Phys. B*, 19(1), 019401, 2010.
[17] Conley A, Vaidyanathan A, Segal C, Heat flux measurements for a GO_2/GH_2 single-element, shear injector , *Journal of Spacecraft and Rockets*, 44(3): 633-639, 2007.
[18] Sozer E, Vaidyanathan A, Segal C, et al, Computational assessment of gaseous reacting flows in single element injector , AIAA Paper No.2009-449.
[19] N. Ierardo, A. Congiunti, C. Bruno, Mixing and Combustion in Supercritical O_2/CH_4 Liquid Rocket Injectors, 42th AIAA Aerospace Sciences Meeting and Exhibit, January 5-8, Reno, Nevada, 2004.
[20] Gao Yushan, Du Zhenggang, Jin Ping, Cai Guobiao, Numerical simulation on the combustion characteristic of shear coaxial GO_2/GCH_4 injector, *Journal of rocket propulsion*, 35(5): 18-23, 2009.

IOP Conf. Series: Earth and Environmental Science **104** (2017) 012003 doi:10.1088/1755-1315/104/1/012003

Turbine Inlet Air Cooling for Industrial and Aero-derivative Gas Turbine in Malaysia Climate

A Nordin[1, a], D A Salim[1, b], M A Othoman[1, c], S N Omar Kamal[1, d], Danny Tam[2, e] and M KY Yusof[2, f]

[1]Mechanical Engineering Department, Politeknik Ungku Omar, Ipoh, Malaysia.
[2]Shinryo (Malaysia) Sdn Bhd, Kuala Lumpur, Malaysia

Corresponding author: [a]adzuieen@puo.edu.my; [b]didi@puo.edu.my; [c]masyraf@puo.edu.my; [d]nshahrul@puo.edu.my; [e]dannytam@shinryo.com.my, [f]kamarul.yusri@shinryo.com.my

Abstract. The performance of a gas turbine is dependent on the ambient temperature. A higher temperature results in a reduction of the gas turbine's power output and an increase in heat rate. The warm and humid climate in Malaysia with its high ambient air temperature has an adverse effect on the performance of gas turbine generators. In this paper, the expected effect of turbine inlet air cooling technology on the annual performance of an aero-derivative gas turbine (GE LM6000PD) is compared against that of an industrial gas turbine (GEFr6B.03) using GT Pro software. This study investigated the annual net energy output and the annual net electrical efficiency of a plant with and without turbine inlet air cooling technology. The results show that the aero-derivative gas turbine responds more favorably to turbine inlet air cooling technology, thereby yielding higher annual net energy output and higher net electrical efficiency when compared to the industrial gas turbine.

1. Introduction

The gas turbine (GT) which was originally used as a jet engine technology and was later transformed into power generation in the early 1990s [1]. The performance of the GT (power output, heat rate, exhaust gas flow and exhaust gas temperature) is greatly dependent on the inlet air temperature to the GT [2]. In a tropical country such as Malaysia the GT its own drawback as its performance is rated at ISO condition with inlet air temperature of 15°C, while the ambient temperature varies between 25°C to 37°C [3]. Power producers de-rate their GT power plants to reflect the higher ambient temperature. It is possible to recover these performance losses by utilizing the air inlet cooling by chilled water. Turbine inlet air cooling (TIAC) technology is designed to chill the inlet air to 15°C and lower temperatures, thereby enhancing the power output and efficiency of the GT.

The GT has been used in Malaysia to generate electricity and the demand for electricity was estimated to increase annually at 4% for the year 2010 to 2014. It is predicted to achieve more than 150000 GWh by 2030 [4].

In the past, studies of the GT inlet air cooling had been undertaken by many authors. All this research has proven that there is an increase in the GT power output when inlet air cooling technology is implemented [5-8]. The majority of industries using inlet air cooling technology are in those part of the world where the climate is tropical and the Middle East as there is a tremendous de-rating factor due to high relative humidity (RH) [9-12].

Drbal et al. [13] in their research stated that ambient air conditions including dry bulb and humidity were important parameters affecting the performance of the GT when it was operated as an open cycle. The researchers reported that the lower air temperatures resulted in a high density of air which led to a high mass flow. As a result, there was a higher output from the turbine since output is proportional to the mass flow of the air inlet.

Jonsson and Yan [14] reported that the implementation of the TIAC technology has increased in recent years due to the increase in the demand for power at a lower cost. The researches made a comparison between two cooling methods (evaporative cooling and mechanical chiller) for the GT's inlet air cooling which was installed at the Khangiran Refinery in Iran. The researches indicated that the inlet air temperature using a mechanical chiller was much lower than that the evaporative cooling method. The increase in thermal efficiency was as low as 7 % to 13.2 %.

De Sa and Al Zubaidi [10] did a study of the SIEMENS Gas Turbine (SGT) 94.2 and SGT 94.3 which were installed at the DEWA Power Station located at AL Aweer in Dubai, UAE. It was reported that at a higher temperature, thermal efficiency and output were lower. For every Kelvin rise of ambient temperature, the resulted showed a 0.1% loss in thermal efficiency. By varying its ambient temperature accordingly to the ISO, both GT's showed an increase of 44 MW of useful power generated.

Hence, the impact of the increased ambient temperature can be dramatic, due to decreased air density and higher fluid temperatures. The compressor absorbs more power when inlet air is hot, leaving a small percentage of total power available for the output [13]. The equations also show that the compressor's specific work is reduced when there is a decrease of the air inlet temperature. Consequently, the GT's efficiency increases.

Mohanty et al. [15] reported that by reducing the ambient temperature to 15°C, the power output of its GT (typical 100MW) showed increases in the range of 8 to 13%. The cooling technique that was applied used an absorption chiller.

In China, Cheng Yang et al. [5] conducted a study using a chiller with Lithium Bromide/water ($LiBr-H_2O$) and also by fogging method to cool down the inlet air temperature. The results showed a superiority in power efficiency when the ambient temperature was in the range of 15°C to 20°C.

Meanwhile in Iran, a research conducted by Farzaneh and Deymi [16] at the Kangiran Gas refinery recorded an increase of 7% to 13.5% of thermal efficiency when there was a drop in the temperature from 46°C to 3°C. This study implemented the evaporative cooling and mechanical chiller methods to cool down the GT's inlet air.

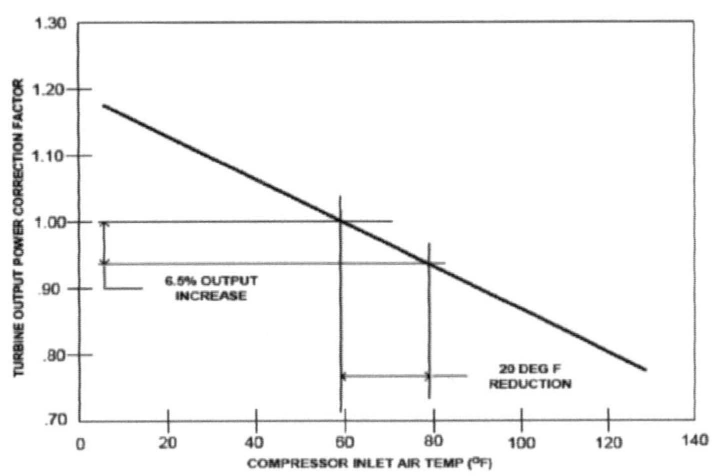

Figure 1. System performance [17]

Figure 1 verifies that the correction factor of the turbine output is dependent on the compressor's inlet air temperature as stated by Sakhaei et al. [17]. This proves that a decrement of 20°F in the inlet air temperature can increase the GT power to approximately 7%. This finding is supported by Komuro et al. [18] as shown in Figure 2 below. The figure shows that by decreasing the inlet air ambient temperature to 15°C, the power output of the GT is 100% efficient. The red array in Figure 2 shows the power which can be recovered as the ambient is at the ISO setting (15°C).

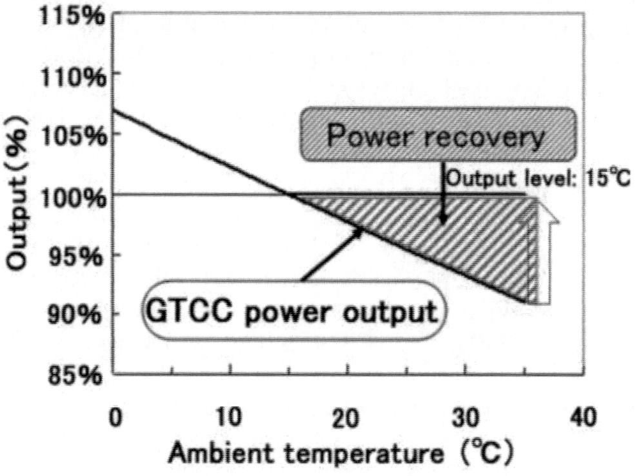

Figure 2. Power output recovery mechanism by the inlet air cooling system [18].

In the current study, the annual performance of different GT (industrial and aero-derivative) systems are evaluated in order to determine the higher power augmentation when TIAC technology is applied.

2. Methodology

The local data of dry bulb, wet bulb and RH for year 2015 from Sitiawan, Perak, Malaysia have been used for this study. The local data were evaluated using the binning method to determine the net power output and overall net electrical efficiency of a plant with and without using TIAC technology.

The data gathered from the meteorology department in the year of 2015 is specified to hourly ambient temperature. Data of wet bulb (WB) temperature, dry bulb (DB) temperature and the relative humidity (RH) were used. Based on these data, a scatter graph of RH versus DB was plotted.

The GT Pro software is used to simulate the annual performance of two gas turbine models which are GE LM6000PD (aero-derivative GT) and GEFr6B.03 (industrial GT) operating with and without Turbine Inlet Air Cooling, in Malaysian climate. For the purpose of annual performance analysis, Sitiawan's weather data is analysed and divided into 27 discrete time periods covering a total of 8756 operating hours. Each time bin indicate the associated operating hours, mean dry bulb and wet bulb temperatures

3. Result and Discussion

The results from the GT Pro software were tabulated for a total of 8,756 operating hours for the year 2015. Table 1 shows the simulation results for aero derivative GT model LM6000PD. The increase in performance is substantially higher for all performance parameters compared to the industrial model of GT.

Table 1. Simulation Results for GTG Model LM6000PD.

Simulation Result		With TIAC	Without TIAC	% Gain from TIAC
Annual Operating Hours	hr	8,756	8,756	
Annual Plant Gross Energy Output	kWh	388,289,450	311,912,549	24.49
Annual Plant Nett Energy Output	kWh	368,740,564	303,543,725	21.48
Annual Plant Fuel LHV Energy Input	kWh	935,284,256	788,545,060	18.61
Annual Plant Gross Heat Rate	kJ/kWh	8,671.43	9,101.15	
Annual Plant Nett Heat Rate	kJ/kWh	9,131.14	9,352.07	
Annual Plant Gross Electrical Efficiency	%	41.52	39.56	4.96
Annual Plant Nett Electrical Efficiency	%	39.43	38.49	2.42

Table 2 shows the simulation results for the industrial GT model GEFr6B.03 operates in an open cycle mode. The results show that there is huge gain in performance with TIAC except for overall plant net electrical efficiency.

Table 2. Simulation Results for GTG Model GEFr6B.03.

Simulation Result		With TIAC	Without TIAC	% Gain from TIAC
Annual Operating Hours	hr	8,756	8,756	
Annual Plant Gross Energy Output	kWh	389,390,997	352,598,974	10.43
Annual Plant Nett Energy Output	kWh	372,551,672	347,342,827	7.26
Annual Plant Fuel LHV Energy Input	kWh	1,163,415,863	1,081,930,437	7.53
Annual Plant Gross Heat Rate	kJ/kWh	10,756.02	11,046.40	
Annual Plant Nett Heat Rate	kJ/kWh	11,242.19	11,213.56	
Annual Plant Gross Electrical Efficiency	%	33.47	32.59	2.70
Annual Plant Nett Electrical Efficiency	%	32.02	32.10	-0.25

Figure 3 shows the annual plant net energy output for both GT models with and without using the TIAC. The introduction of the TIAC has improved the net power output by 7.26 % for the industrial model and 21.48 % for the aero- derivative model respectively.

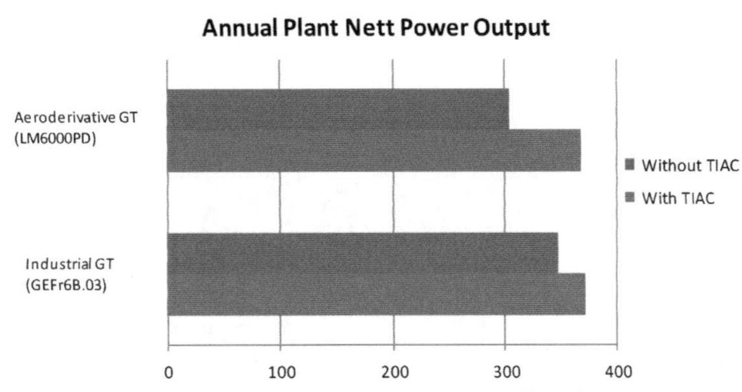

Figure 3. Annual Plant Nett Energy Output

As shown in Figure 4, with TIAC the annual plant net heat rate for the GE LM6000PD is expected to decrease by 2.36% while for the GEFr6B.03, it is expected to increase by 0.25%. heat rate is the amount of fuel energy it heat input required to generate a kilowatt hour and deliver it to the transmission line [19].

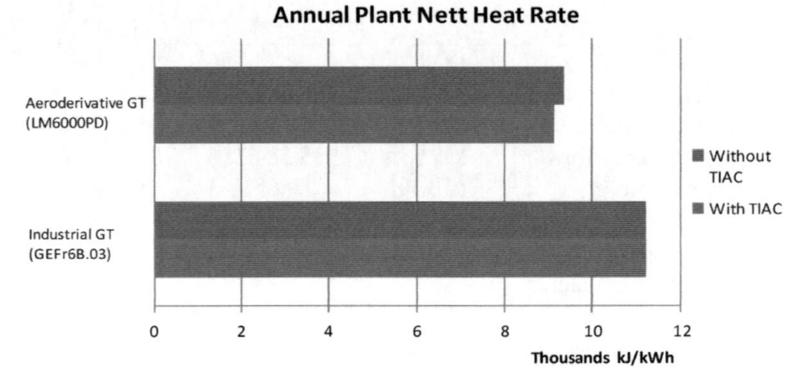

Figure 4. Annual Plant Nett Heat Rate

Figure 5 illustrates the annual plant net electrical efficiency for both GT models. The implementation of the TIAC technology has improved the net electrical efficiency by 2.42% for the aero-derivative GT. However the overall plant net plant electrical efficiency is marginally reduced by 0.25%.

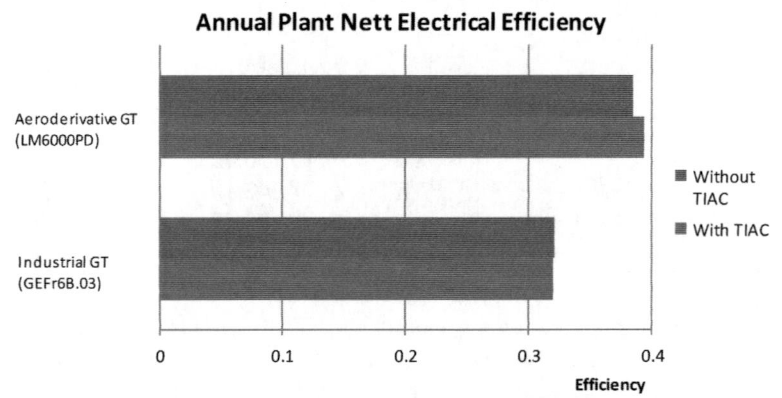

Figure 5. Annual Plant Nett Electrical Efficiency

From the result obtained, it can be concluded that both GT models are responsive to the difference of turbine inlet air temperatures. The GE LM600PD, being an aero-derivative machine, is highly responsive to the inlet air temperature. By contrast, the GEFr6B.03, being an industrial gas turbine, is moderately responsive to the inlet air temperature.

With the inlet air cooled to 12°C, the annual net power output and net electrical efficiency can be improved when the parasitic load of the chiller system for turbine inlet air cooling is taken into account

4. Conclusion

The aero-derivative heritage of the GE LM600PD enables the machine to respond more favourably to TIAC power augmentation technology, thereby yielding higher annual net energy output and higher annual plant nett electrical efficiency when compared to the industrial GEFr6B.03. Hence it may be worthwhile for plants with aero-derivative GT installations to consider implementing TIAC technology to reap the potential benefits of increased power output and improved net electrical efficiency. Plants with industrial GTs can expect increased power output with TIAC albeit with potentially lower net electrical efficiency.

Acknowledgement
The authors gratefully acknowledge the full financial and expertise support provided by SHINRYO (MALAYSIA) SDN BHD.

References

[1] Breeze, P., Gas-turbine Power Generation. 2016: Academic Press.
[2] Al-Ansary, H.A., J.A. Orfi, and M.E. Ali, Impact of the use of a hybrid turbine inlet air cooling system in arid climates. Energy conversion and management, 2013. 75: p. 214-223.
[3] Annual Climatic Conditions of Sitiawan, P., J.M. Malaysia, Editor 2015.
[4] Behrouzi, F., et al., Global renewable energy and its potential in Malaysia: A review of Hydrokinetic t*urbine technology*. Renewable and Sustainable Energy Reviews, 2016. **62**: p. 1270-1281.
[5] Yang, C., Z. Yang, and R. Cai, *Analytical method for evaluation of gas turbine inlet air cooling in combined cycle power plant.* Applied Energy, 2009. **86**(6): p. 848-856.
[6] Kakaras, E., A. Doukelis, and S. Karellas, *Compressor intake-air cooling in gas turbine plants.* Energy, 2004. **29**(12): p. 2347-2358.
[7] De Pascale, A., F. Melino, and M. Morini, *Analysis of inlet air cooling for IGCC power augmentation.* Energy Procedia, 2014. **45**: p. 1265-1274.
[8] Chacartegui, R., et al., *Analysis of combustion turbine inlet air cooling systems applied to an operating cogeneration power plant.* Energy conversion and management, 2008. **49**(8): p. 2130-2141.
[9] Dawoud, B., Y. Zurigat, and J. Bortmany, *Thermodynamic assessment of power requirements and impact of different gas-turbine inlet air cooling techniques at two different locations in Oman.* Applied Thermal Engineering, 2005. **25**(11): p. 1579-1598.
[10] De Sa, A. and S. Al Zubaidy, *Gas turbine performance at varying ambient temperature.* Applied Thermal Engineering, 2011. **31**(14): p. 2735-2739.
[11] Sebzali, M., B. Ameer, and H. Hussain, *Comparison of energy performance and economics of chilled water thermal storage and conventional air-conditioning systems.* Energy and Buildings, 2014. **69**: p. 237-250.
[12] Ahmadi, P. and I. Dincer, *Thermodynamic and exergoenvironmental analyses, and multi-objective optimization of a gas turbine power plant.* Applied Thermal Engineering, 2011. **31**(14): p. 2529-2540.
[13] Drbal, L., K. Westra, and P. Boston, *Power plant engineering.* 2012: Springer Science & Business Media.
[14] Jonsson, M. and J. Yan, *Humidified gas turbines—a review of proposed and implemented cycles.* Energy, 2005. **30**(7): p. 1013-1078.
[15] Mohanty, B. and G. Paloso, *Enhancing gas turbine performance by intake air cooling using an absorption chiller.* Heat Recovery Systems and CHP, 1995. **15**(1): p. 41-50.
[16] Farzaneh-Gord, M. and M. Deymi-Dashtebayaz, *Effect of various inlet air cooling methods on gas turbine performance.* Energy, 2011. **36**(2): p. 1196-1205.
[17] Sakhaei, S.A. and M. Safari, *Study and Comparison of Inlet Air Cooling Technique of Gas Turbines and Their Effects on Increase of the Efficiency and Outlet Power.* International Journal of materials, mechanics and manufacturing, 2014. **2**(4): p. 329-334.
[18] Komuro, T., E. Ito, and T. Sonoda, *Power output augmentation of gas turbine combined cycle by inlet-air cooling system of chiller type under high ambient temperature.* Mitsubishi Heavy Industries Technical Review, 2010. **47**(4).
[19] Black, V., et al., *Power plant engineering.* International Thomson, New York, 1996.

IOP Conf. Series: Earth and Environmental Science **104** (2017) 012004 doi:10.1088/1755-1315/104/1/012004

Joint Planning Of Energy Storage and Transmission Considering Wind-Storage Combined System and Demand Side Response

Y Huang[1], B Z Liu[1], K Y Wang[1] and X Ai[1]

[1]State Key Laboratory of Alternate Electrical Power System with Renewable Energy Sources, North China Electric Power University, Beijing 102206, China

1765506332@qq.com, bzliu@ncepu.edu.cn, wky_ncepu@163.com, aixin@ncepu.edu.cn

Abstract. In response to the new requirements of the operation mode of wind-storage combined system and demand side response for transmission network planning, this paper presents a joint planning of energy storage and transmission considering wind-storage combined system and demand side response. Firstly, the charge-discharge strategy of energy storage system equipped at the outlet of wind farm and demand side response strategy are analysed to achieve the best comprehensive benefits through the coordination of the two. Secondly, in the general transmission network planning model with wind power, both energy storage cost and demand side response cost are added to the objective function. Not only energy storage operation constraints and but also demand side response constraints are introduced into the constraint condition. Based on the classical formulation of TEP, a new formulation is developed considering the simultaneous addition of the charge-discharge strategy of energy storage system equipped at the outlet of the wind farm and demand side response strategy, which belongs to a typical mixed integer linear programming model that can be solved by mature optimization software. The case study based on the Garver-6 bus system shows that the validity of the proposed model is verified by comparison with general transmission network planning model. Furthermore, the results demonstrate that the joint planning model can gain more economic benefits through setting up different cases.

1. Introduction

In recent years, structure and operation mode of modern power system have changed greatly with the development of renewable energy, energy storage technology and load side management [1-4]. The source side or the load side change will have a substantial impact on the transmission network planning. That's because it is based on power and load forecasting data. Therefore, it is necessary to take full account of the impacts of wind farms and energy storage systems as well as load side management on planning.

Now, there is the discussion on the combination of energy storage planning and transmission network planning [5-6] and the combination of demand side response and transmission network planning [7-10]. However, there is not much research on the transmission network planning considering the two simultaneously. For example, literature [5] establishes a two layers planning model of a transmission system considering only the operation of a wind-storage combined system. A transmission planning model only based on demand side response mechanism is established in the literature [7].

Therefore, this paper proposes a joint planning of energy storage and transmission considering wind-storage combined system and demand side response. Energy storage operation constraints and demand side response constraints are introduced. On this basis comprehensive resources planning is realized. The example system shows that the model is comparable, and the joint planning model can achieve more significant economic benefits.

2. Operation strategy analysis

2.1. Operation strategy of wind-storage combined system

To alleviate the pressure on the system regulating capacity and effectively reduce wind power curtailment, energy storage system equipped at the outlet of wind farm can be used to smooth stochastic fluctuation of wind farm output power. Combined with energy storage characteristics that frequent charge-discharge can shorten the cycle life, the operating strategies are proposed: in the planning model, the constraints of energy storage operation and daily charge-discharge are introduced. In this way energy storage configuration capacity is optimized to meet the requirement of wind wave rate and to maximize utilization ratio of wind resources.

2.2. Demand side response strategy

Demand side response is that electric power department encourages users to change power consumption mode through administrative and economic means, which can achieve peak shaving, valley filling, load transfer and relieve system regulating pressure. The numbers of new power plants and new transmission lines can be reduced on the planning layer [4]. Therefore, this paper proposes the following demand side response strategy only considering interruptible load and incentive load response: the demand side response strategy is introduced in the planning transmission system with wind power. Under the system security constraints, when the load is high, the wind power supply is less or the system regulation capacity is insufficient, the users who sign interruptible load protocols are guided to reduce power consumption. When the load is low, the wind power supply is adequate and the system regulation capacity is insufficient, the users who sign incentive load protocols are guided to increase power consumption. Otherwise, there will be wind power curtailment. In this way, the costs of transmission line and wind power curtailment are reduced simultaneously.

3. Mathematical model

3.1. Objective function

The objective function of the model consists of four parts: investment cost of transmission line f_{line}, investment cost of energy storage system f_{ess}, cost of wind power curtailment f_{wind} and compensation cost of demand side response f_{dsr}. Since the model of this paper takes the time dimension into account, all the cost functions in the objective function are translated into annual investments.

$$\min f = f_{line} + f_{ess} + f_{wind} + f_{dsr} \tag{1}$$

$$f_{line} = \frac{i(1+i)^n}{(1+i)^n - 1} \sum_{ij\in\Omega_l^p} c_{ij} n_{ij}^p \tag{2}$$

$$f_{ess} = (1+r_{op} + r_{ma} + r_{sc}) r_{de} \sum_{k\in\Omega_{es}} (c_p p_{es,k}^r + c_e e_{es,k}^r) \tag{3}$$

$$f_{wind} = \sum_{j\in\Omega_w} \sum_{s\in\Gamma} d_s \sum_{t=1}^{T} c_w (p_{w,t,s,j}^{\max} - p_{w,t,s,j}) \tag{4}$$

$$f_{dsr}=\sum_{s\in\Gamma} d_s \sum_{t=1}^{T}(c_t^+ \sum_{m\in\Omega_L^+} \Delta p_{L,t,s,m}^+ + c_t^- \sum_{n\in\Omega_L^-} \Delta p_{L,t,s,n}^-) \tag{5}$$

In equation (2), i is the discount rate and n is service life cycle. ij represents the branch of the head and the tail nodes numbered i and j, c_{ij} is the cost of single branch ij, and n_{ij}^p is the extension number of the p-th candidate branch ij. Ω_l^p is the set of candidate branches. In equation (3), r_{op} , r_{ma} and r_{sc} are cost conversion coefficients of the operation, maintenance and disposal of the energy storage system. c_p and c_e are the unit power and unit capacity costs of energy storage system. $P_{es,k}^r$ and $e_{es,k}^r$ are the rated power and rated capacity of the energy storage system allocated at the node k . Ω_{es} is the set of energy storage nodes to be configured. In equation (4), Ω_w is the set of wind farms. The letter 's' stands for a typical day, the number of days of which is d_s. T is the number of periods per typical day, Γ is the set of typical days, and c_w is cost of unit wind power curtailment. $p_{w,t,s,j}^{max}$ and $p_{w,t,s,j}$ are the maximum output and actual output of wind farm j on typical day s within hour t .In equation (5), c_t^+ and c_t^- are unit compensation costs of interruptible load and incentive load within hour t. Ω_L^+ and Ω_L^- are the sets of interruptible load and incentive load. $\Delta p_{L,t,s,m}^+$ is the increase of incentive load on typical day s within hours t for node m and $\Delta p_{L,t,s,n}^-$ is the decrement of interruptible load on typical day s within hour t for node m.

3.2. Constraints

The primary constraints of the model include the system DC power flow constraints, conventional power plant operation constraints, wind farm operation constraints, energy storage system operation constraints, demand side response constraints and system standby constraints.

3.2.1. System DC power flow constraints

$$\boldsymbol{P}_{g,t,s} + \boldsymbol{P}_{ws,t,s} - \boldsymbol{A}^o \boldsymbol{P}_{ij,t,s}^o - \boldsymbol{A}^p \boldsymbol{P}_{ij,t,s}^p = \boldsymbol{P}_{l,t,s} + \boldsymbol{\Delta P}_{L,t,s}^+ - \boldsymbol{\Delta P}_{L,t,s}^- \tag{6}$$

$$\begin{cases} p_{ij,t,s}^o - b_{ij} n_{ij}^o (\theta_{i,t,s} - \theta_{j,t,s}) = 0, \left| p_{ij,t,s}^o \right| \le n_{ij}^o p_{ij}^{\max}, \ \forall (i,j) \in \Omega_l^o; \\ \left| p_{ij,t,s}^p - b_{ij} (\theta_{i,t,s} - \theta_{j,t,s}) \right| \le M(1 - n_{ij}^p), \left| p_{ij,t,s}^p \right| \le n_{ij}^p p_{ij}^{\max}, \ \forall (i,j) \in \Omega_l^p; \\ p \in (1, 2, \cdots, n_{ij}^{\max}), \theta_{ref,t,s} = 0. \end{cases} \tag{7}$$

In equation (6), $\boldsymbol{P}_{g,t,s}$, $\boldsymbol{P}_{ws,t,s}$ and $\boldsymbol{P}_{l,t,s}$ stand for output vectors of conventional power plant, output vector of wind-storage combined system and load active demand vector at each node on typical day s within hour t, respectively. $\boldsymbol{A}^o$ and $\boldsymbol{A}^p$ stand for node-branch incidence matrices of existing and candidate lines. $\boldsymbol{P}_{ij,t,s}^o$ and $\boldsymbol{P}_{ij,t,s}^p$ are branch active power vectors of existing and candidate lines on typical day s within hour t, respectively.In equation (7), $p_{ij,t,s}^o$ and $p_{ij,t,s}^p$ are active power of existing and candidate branch ij on typical day s within hour t, respectively. b_{ij} is admittance of branch ij. n_{ij}^o is the number of existing line ij, and n_{ij}^{max} is the maximum optional number of candidate line ij. $\theta_{i,t,s}$, $\theta_{j,t,s}$ and $\theta_{ref,t,s}$ are voltage phase angles of bus i, bus j and slack bus on typical day s within hour t, respectively. p_{ij}^{max} is the maximum active power limit of branch ij.

3.2.2. Conventional power plant operation constraints

$$p_{g,i}^{\min} \le p_{g,t,s,i} \le p_{g,i}^{\max}, \Delta p_{g,i}^{\min} \le p_{g,t,s,i} - p_{g,t-1,s,i} \le \Delta p_{g,i}^{\max} \tag{8}$$

Where max and min stand for high/low limit of corresponding variables. $p_{g,t,s,i}$ and $p_{g,t-1,s,i}$ are active power of conventional power plant i on typical day s within hour t and hour t-1, respectively. $\Delta p_{g,i}^{min}$ and $\Delta p_{g,i}^{max}$ stand for equivalent landslide and climbing rate of conventional power plant i, respectively.

3.2.3. Wind farm operation constraints

$$0 \le p_{w,t,s,j} \le p_{w,t,s,j}^{\max}, p_{ws,t,s,j} = p_{w,t,s,j} - p_{es,t,s,j}, \left| 1 - \frac{p_{ws,t,s,j}}{p_{ws,t-1,s,j}} \right| \le \gamma_{ws} \tag{9}$$

Where $p_{w,t,s,j}$ and $p_{es,t,s,j}$ are actual output of wind farm *j* and charge-discharge power of energy storage system *j* on typical day *s* within hour *t*, respectively. $p_{ws,t,s,j}$ and $p_{ws,t-1,s,j}$ are output of wind-storage combined system *j* on typical day *s* within hour *t* and hour *t*-1, respectively. γ_{ws} is the fluctuation limit of output of wind-storage combined system.

3.2.4. Energy storage system operation constraints

$$0 \le e_{es,t,s,k} \le e^{r}_{es,k}, -p^{r}_{es,k} \le p_{es,t,s,k} \le p^{r}_{es,k}, p^{r,\min}_{es,k} \le p^{r}_{es,k} \le p^{r,\max}_{es,k} \quad (10)$$

$$e_{es,1,s,k} = e_{es,T,s,k} = 0, e_{es,1,s,k} = e_{es,T,s,k} + p_{es,T,s,k}\Delta t, e_{es,t,s,k} = e_{es,t-1,s,k} + p_{es,t-1,s,k}\Delta t \quad (11)$$

In equation (10), $e_{es,t,s,k}$ is capacity of energy storage system *j* on typical day *s* within hour *t*. In equation (11), $e_{es,1,s,k}$, $e_{es,T,s,k}$ and $e_{es,t,s,k}$ are storage energy of energy storage system *k* on typical day *s* within hour 1, T and *t*, respectively. $p_{es,T,s,k}$ and $p_{es,t-1,s,k}$ are charge-discharge power of energy storage system *k* on typical day *s* within hour *T* and *t*-1, respectively. Δt is time interval,1h.

$$flag_{t,s,k} = sign(p_{es,t,s,k}) = \begin{cases} 1, p_{es,t,s,k} > 0 \\ 0, p_{es,t,s,k} = 0 \\ -1, p_{es,t,s,k} < 0 \end{cases} \quad (12)$$

$$\begin{cases} flag_{t,s,k} \le -\frac{1}{2}u^{d}_{t,s,k} + (1-u^{d}_{t,s,k})M, flag_{t,s,k} \ge -\frac{1}{2}u^{c}_{t,s,k} + (1-u^{c}_{t,s,k})M, \\ u^{c}_{t,s,k} + u^{d}_{t,s,k} = 1, u_{t,s,k} = u^{c}_{t,s,k} - u^{d}_{t,s,k}, \frac{1}{2}\sum_{t=2}^{T}\left|u_{t,s,k} - u_{t-1,s,k}\right| \le N^{\max}_{dc} \end{cases} \quad (13)$$

In equation (12), $flag_{t,s,k}$ is variables representing charging and discharging state of energy storage, and possible values are -1, 0 and 1.In equation (13), $u^{d}_{t,s,k}$ and $u^{c}_{t,s,k}$ are 0-1 variables representing charging and discharging state of energy storage. $u_{t,s,k}$ is the variables representing charging and discharging state of energy storage, and possible values are -1 and 1. *M* is a big number. N^{max}_{dc} is the maximum charge-discharge times of an energy storage system on each typical day.

3.2.5. Demand side response constraints

$$u^{+}_{L,t,s,m} + u^{-}_{L,t,s,m} \le 1, 0 \le \Delta p^{+}_{L,t,s,m} \le \Delta p^{+,\max}_{L,t,s,m} u^{+}_{L,t,s,m}, 0 \le \Delta p^{-}_{L,t,s,m} \le \Delta p^{-,\max}_{L,t,s,m} u^{-}_{L,t,s,m} \quad (14)$$

$$\sum_{t}^{t+T^{+}_{\max}} u^{+}_{L,t,s,m} \le T^{+}_{\max}, t = 1,2,\cdots,T - T^{+}_{\max}; \sum_{t}^{t+T^{-}_{\max}} u^{-}_{L,t,s,m} \le T^{-}_{\max}, t = 1,2,\cdots,T - T^{-}_{\max} \quad (15)$$

In equation (14), $u^{+}_{L,t,s,m}$ and $u^{-}_{L,t,s,m}$ are 0-1 variables representing the operation state of incentive load and interruptible load. In equation (15), T^{+}_{max} and T^{-}_{max} are the maximum continuous time of operation of incentive load and interruptible load.

3.2.6. System standby constraints

$$\begin{cases} \sum_{i} p^{\max}_{g,t,s,i} + \sum_{j} p_{ws,t,s,j} \ge \sum_{k} p_{L,t,s,k} - \sum_{m} \Delta p^{-}_{L,t,s,m} + R_{u,t}, \\ \sum_{i} p^{\max}_{g,t,s,i} + \sum_{j} p_{ws,t,s,j} \ge \sum_{k} p_{L,t,s,k} + \sum_{n} \Delta p^{+}_{L,t,s,n} - R_{d,t} \end{cases} \quad (16)$$

Where $R_{u,t}$ and $R_{d,t}$ are the positive and negative rotating reserve capacity limit within hours *t*.

3.3. Model solving

The model established in this paper belongs to a typical mixed integer linear programming model. The objective function consists of equations (1)-(5), the constraints consist of equations (6)-(16) and decision variables consist of integer and continuous types. Integer variables are n^{p}_{ij} , $u_{t,s,k}$, $u^{+}_{L,t,s,m}$, $u^{-}_{L,t,s,m}$ and so on, which determine transmission line planning scheme, charging and discharging state of energy storage system and participation status of demand side response load, respectively. Continuous variables are $p^{o}_{ij,t,s}$, $p^{p}_{ij,t,s}$, $p_{g,t,s}$, $p_{w,t,s}$, $\Delta p^{+}_{L,t,s}$, $\Delta p^{-}_{L,t,s}$, $p_{es,t,s}$, $e_{es,t,s}$, $p^{r}_{es,k}$ and $\theta_{i,t,s}$. The case has been implemented on a personal computer, using GUROBI 7.0.2 under MATLAB.

4. Case study

4.1. Example system settings

This paper takes the Garver-6 bus system widely used by transmission network planning researchers as an example system [4]. System parameters are available in the literature [4] and original topology is shown in figure 1. Suppose a new power plant bus is added and total maximum load forecast is increased to 760MW. There are 15 possible expansion corridors, and the maximum number of lines allowed for each corridor is set to 3. A wind farm with a capacity of 240MW is added at bus 3. It is assumed that all load nodes in the system can participate in incentive response, and nodes 2, 4 and 5 participate in interruptible response. Line cost is 100 million RMB/km. Cost of unit wind power curtailment is 0.61RMB/kWh [5]. The compressed-air system is selected as the energy storage system, and cost parameters [11-12] of which are shown in Table 1. The time-sharing compensation price of demand side response is shown in Table 2.

Table 1. The cost parameters of energy storage system.

r_{op}	r_{de}	c_p(RMB/kW)	c_e(RMB/kWh)
0.01	0.05	200	4000

Table 2. Compensation price.

Compensation price (RMB/kWh)	Rush hours 9:00-14:00 18:00-22:00	Slack hours 0:00-7:00	Other hours 7:00-9:00 14:00-18:00 22:00-24:00
c_t^{+}	0.1	0.5	0.3
c_t^{-}	0.5	0.1	0.3

4.2. Planning result analysis

4.2.1. Validity comparison In order to make the result of this model comparable, this section does not consider the wind power integration, energy storage integration, and demand side response strategy without considering the fluctuations of original power and load. The planning result is obtained by solving model, which is shown in figure 2. The result is in good agreement with the results of literature [4] by means of stepwise inverse method and literature [13] using Benders decomposition method, which validates the validity of the model of this paper.

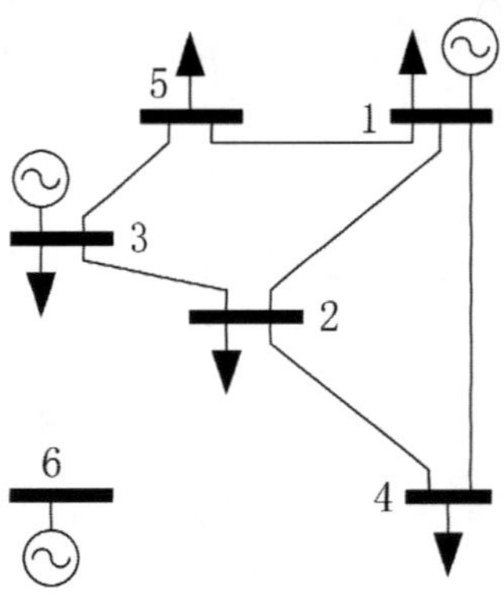

Figure 1. Initial network topology.

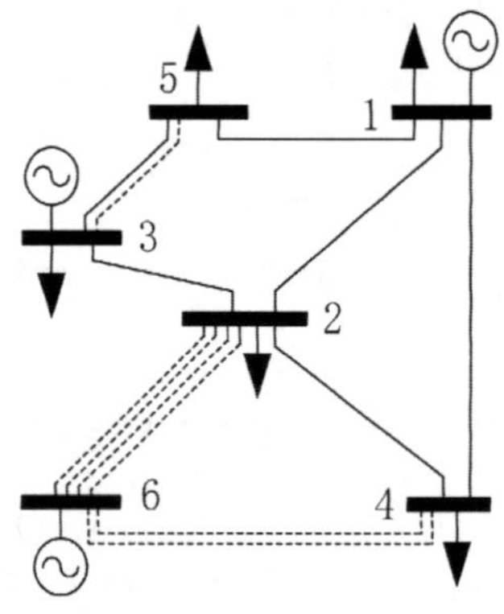

Figure 2. Optimal expansion plan result.

4.2.2. Comparison of results of different planning cases. To reflect advantages of synthetically considering the combination of energy storage system and demand side response as well as transmission network planning with wind power, four different cases are set up. Case 1: neither the energy storage system integration nor the demand side response strategy is considered. Case 2: only demand side response strategy is considered. Case 3: only energy storage system is considered. Case 4: both energy storage system integration and demand side response strategies are considered. The planning results are obtained by solving models: planning schemes under different cases are shown in table 3, and planning costs under different cases are shown in table 4.

Analyses are conducted on the planning results in tables 3 and 4:

Compared with case 1, it can be seen that line cost is reduced by 3.41 million RMB and cost of wind power curtailment is reduced by 6.05 million RMB in case 2, meanwhile cost of wind power curtailment is reduced by 48.55 million RMB in case 3. By comparing cases 2 and 3, it can be seen that cost of wind power curtailment drops to 0, and the annual total investment cost reaches the minimum. Results demonstrate that demand side response strategy can reduce peak load for load curve, thereby reducing the investment costs of transmission lines. An energy storage system is added to the side of the wind farm can not only stabilize wind generation fluctuation but also improve system regulation capability. In addition, considering the influence of energy storage system and demand side response on planning, wind power capacity is improved with economic benefits.

Table 3. Planning schemes under different cases.

Case	Route planning scheme	Stored energy planning power
case1	2-6(3) 2-5(1) 3-5(1) 4-6(2)	0
case2	2-6(3) 3-5(1) 4-6(2)	0
case3	2-6(3) 2-5(1) 3-5(1) 4-6(2)	44.04MW
case4	2-6(3) 3-5(1) 4-6(2)	41.66MW

Table 4. Planning costs under different cases. unit: Million RMB

Case	f_{line}	f_{ess}	f_{wind}	f_{dsr}	f
case1	22.36	0	51.96	0	74.32
case2	18.95	0	45.91	4.60	69.46
case3	22.36	10.67	3.41	0	36.44
case4	18.95	10.10	0	2.59	31.64

Figure 3 shows energy storage charge-discharge power curve and SOC curve under case 4. Figure 4 presents fluctuation rate of output of integrated system under case 4.

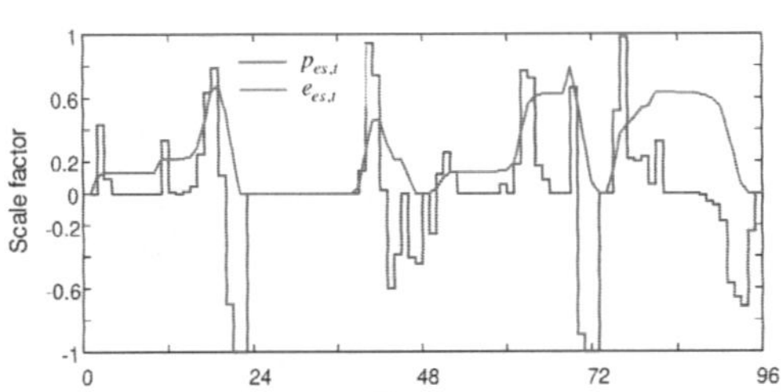

Figure 3. Energy storage charge discharge power curve and SOC curve.

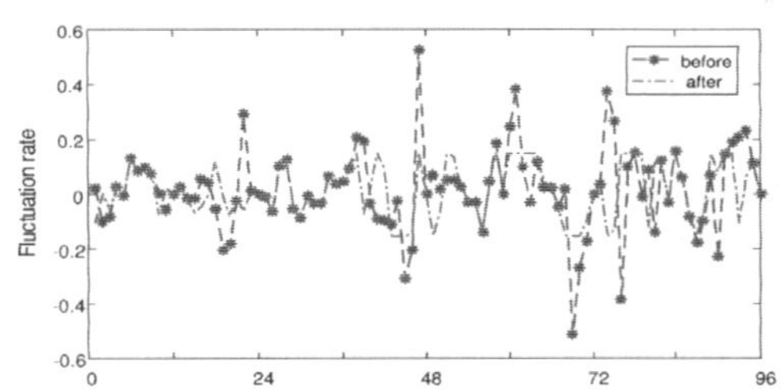

Figure 4. Fluctuation rate of output of integrated system.

5. Conclusion

The combined operation mode of wind farm and energy storage changes the output of the source, and demand side response patterns change the load side demand characteristics. Under the new situation, the transmission network planning faces the double challenges due to changes at source side and load

IOP Conf. Series: Earth and Environmental Science **104** (2017) 012004 doi:10.1088/1755-1315/104/1/012004

side. Therefore, in the transmission network planning, it is necessary to comprehensively consider wind-storage combined system and demand side response.

The case study based on the Garver 6-bus system shows: compared with the model that takes only one factor (energy storage system equipped at the outlet of wind farm or demand side response strategy) into account, the proposed model that considers the mutual response of energy storage and demand side response can enhance the capacity of wind power utilization, optimize the energy storage configuration capacity and line planning scheme as well as bring the greatest economic benefits on the basis of ensuring the safe and reliable operation of the system.

Acknowledgments

This work is supported by the National Key R&D Program of China (2016YFB0900500). We would like to express our most sincere appreciation.

References

[1] Shu Y B, Zhang Z G, Guo J B and Zhang Z L 2017 Study on key factors and solution of renewable energy accommodation *Proceedings of the CSEE* **37** 1-8

[2] Yuan X M, Cheng S J and Wen J Y 2013 Prospects analysis of energy storage application in grid integration of large-scale wind power *Automation of Electric Power Systems* **37** 14-18

[3] Zhu L, Yan Z, Yang X and Fu Y 2014 Integrated resources planning in microgrid based on modeling demand response *Proceedings of the CSEE* **34** 2621-28

[4] Sun H B 1996 *Electric power network planning* (Chongqing: Chongqing University Press) chapter 5 pp 103-106

[5] Zheng J, Wen F S, Li L, Wang K and Gao C 2013Transmission system expansion planning considering combined operation of wind farms and energy storage systems *Automation of Electric Power Systems* **37** 135-142

[6] Liu S T, Wang Y and Guo H M 2013 Transmission system planning and benefit analysis with wind and storage integration *Shaan xi Electric Power* **41** 26-29

[7] Zeng M, Lv C Q, Qiu L Q and Tian K 2011 A demand-side response-based transmission planning model with grid-connected wind farms *Power System Technology* **35** 129-134

[8] Zheng J, Wen F Q, Zhou M L, Xu L Z, Liu W D and Liang L 2014 Transmission system planning for power systems with wind generators considering demand side responses *Journal of North China Electric Power University* **41** 42-48

[9] Zeng M and Lv Q 2016 Coordinated planning model of wind power and transmission grid investment considering demand side response *Shaan xi Electric Power* **44** 1-4

[10] Zeng M, Yang Y Q, Xiang H W, Wang L H and Sun J H 2016 Robust optimization planning model of power system considering demand response *Automation of Electric Power Systems* **40** 137-145

[11] Wang K Y 2016 Risk measure based approach for CVPP operation *North China Power University*

[12] Chacra F A, Bastard P and Fleury G. 2005 Energy storage associated to wind farms in a two-tariff structure bidding tender IEEE 1-5.

[13] Macrae C A G, Ernst A T and Ozlen M 2016 A Benders decomposition approach to transmission expansion planning considering energy storage *Energy* **112** 795-803

Optimal Output of Distributed Generation Based On Complex Power Increment

D Wu[1, *] and H Bao[1]

[1] School of Electrical & Electronic Engineering, North China Electric Power University, Beijing 102206, China

[*]13366306121@163.com

Abstract. In order to meet the growing demand for electricity and improve the cleanliness of power generation, new energy generation, represented by wind power generation, photovoltaic power generation, etc has been widely used.The new energy power generation access to distribution network in the form of distributed generation, consumed by local load. However, with the increase of the scale of distribution generation access to the network, the optimization of its power output is becoming more and more prominent, which needs further study. Classical optimization methods often use extended sensitivity method to obtain the relationship between different power generators, but ignore the coupling parameter between nodes makes the results are not accurate; heuristic algorithm also has defects such as slow calculation speed, uncertain outcomes. This article proposes a method called complex power increment, the essence of this method is the analysis of the power grid under steady power flow. After analyzing the results we can obtain the complex scaling function equation between the power supplies, the coefficient of the equation is based on the impedance parameter of the network, so the description of the relation of variables to the coefficients is more precise Thus, the method can accurately describe the power increment relationship, and can obtain the power optimization scheme more accurately and quickly than the extended sensitivity method and heuristic method.

1. Introduction

Energy shortage and environmental problems are the bottleneck of the development of energy industry. In the electric power industry, the bottleneck behave as the uneven distribution of power and load, and the environmentally unfriendly power generation process. In order to balance the difference of power and load of China's eastern region and western region, help save energy and reduce emission, in the "13th Five-Year of power development plan" the National Energy Bureau stressed the priority layout of clean energy, to ensure the completion of the hard target that non fossil energy consumption reach to 15%[1].Due to policy incentives and technological advances in power generation, the new energy generation represented by wind power generation, photovoltaic power generation and energy storage system has been introduced into distribution network in the form of distributed power supply[2].The flexibility of power generation, the cleanness of process and other characteristics, lead to the scale of distributed power generation becoming more and more large, many problems have emerged, including grid connected mode, optimized configuration and optimized scheduling[3].

When the scale of distributed generation connected to the grid is small, administrators apply the management mode called "fit-and-forget"[4], with the continuous expansion of the scale of distributed generation and the increase of adjustability of distributed generation, it is feasible and necessary to optimize the output planning. At present, the method of distributed generation optimization is divided into heuristic algorithm and classical mathematical optimization algorithm. The optimization target can be divided into two aspects: the highest comprehensive profitability and the highest profitability of Power Grid Corp. Document [5-8] uses heuristic algorithm to solve their own objective function, including genetic algorithm, particle swarm optimization algorithm, difference algorithm and so on. In the application of heuristic algorithm, each article has improved the algorithm for its own problems, but still cannot avoid the shortcomings such as slow speed of calculation and unstable results. In document [9, 10], the classical mathematical optimization method is adopted to optimize the objective function, including interior point method and two sequence programming method. The core of the classical method is the calculation of extended sensitivity. In classical mathematical optimization method, the calculation of the extended sensitivity ignores the coupling relation among the node variables, so the result is not accurate.

2. Analysis of extended sensitivity and complex power increment

With the progress of power generation technology, distributed power supply has a certain capacity of reactive power regulation [11, 12], in the limited power range, the power output can be adjusted freely. Based on this, the distributed power is considered as the power source with adjustable active and reactive power. There is an equation in the steady state power flow of power grid:

$$S_0 = \sum_{i=1}^{l} S_{Li} - \sum_{i=1}^{q} S_{DGi} + S_{loss} \tag{1}$$

Among them, S_0 is the injection power of the system node, and is regarded as the slack node in the text, i is the serial number of the load power, and the total quantity of the load is expressed by l, S_{DGi} is the distributed generation output power, the total quantity expressed by q, S_{loss} is the current loss power of the network. According to the equation, the key point is to analyze the incremental relationship between the network loss and the power supply in order to extract the incremental relationship between S_0 and S_{DGi}.

2.1. Grid loss analysis of extended sensitivity

The grid loss analysis of extended sensitivity is essentially to find out the relation among node power and grid loss power, node voltage value and phase angle respectively. After dividing, the incremental relationship between grid loss and power supply can be obtained. The concrete deduction process is as follows.

$$P_{loss} = \sum_{i=1}^{n} U_i * \sum_{j=1}^{n} U_j * (G_{ij} * \cos q_{ij} + B_{ij} * \sin q_{ij}) \tag{2}$$

$$Q_{loss} = \sum_{i=1}^{n} U_i * \sum_{j=1}^{n} U_j * (G_{ij} * \sin \theta_{ij} - B_{ij} * \cos \theta_{ij}) \tag{3}$$

$$P_i = U_i * \sum_{j=1}^{n} U_j * (G_{ij} * \cos \theta_{ij} + B_{ij} * \sin \theta_{ij}) \tag{4}$$

$$Q_i = U_i * \sum_{j=1}^{n} U_j * (G_{ij} * \sin \theta_{ij} - B_{ij} * \cos \theta_{ij}) \tag{5}$$

According to formula (2) ~ (5), the formula is deduced as follows:

$$\frac{dP_{loss}}{dP_i} = \sum_{m=1}^{n-1} \frac{dP_{loss}}{dU_m} * \frac{1}{dP_i / dU_m} + \sum_{m=1}^{n-1} \frac{dP_{loss}}{d\theta_m} * \frac{1}{dP_i / d\theta_m} \tag{6}$$

$$\frac{dP_{loss}}{dQ_i} = \sum_{m=1}^{n-1} \frac{dP_{loss}}{dU_m} * \frac{1}{dQ_i / dU_m} + \sum_{m=1}^{n-1} \frac{dP_{loss}}{d\theta_m} * \frac{1}{dQ_i / d\theta_m} \tag{7}$$

$$\frac{dQ_{loss}}{dP_i} = \sum_{m=1}^{n-1} \frac{dQ_{loss}}{dU_m} * \frac{1}{dP_i / dU_m} + \sum_{m=1}^{n-1} \frac{dQ_{loss}}{d\theta_m} * \frac{1}{dP_i / d\theta_m} \tag{8}$$

$$\frac{dQ_{loss}}{dQ_i} = \sum_{m=1}^{n-1} \frac{dQ_{loss}}{dU_m} * \frac{1}{dQ_i / dU_m} + \sum_{m=1}^{n-1} \frac{dQ_{loss}}{d\theta_m} * \frac{1}{dQ_i / d\theta_m} \tag{9}$$

In the process of calculating the extended sensitivity, the coupling relation between the voltage value and the phase angle of different nodes is neglected, which results in inaccurate results.

2.2. *Grid loss analysis of complex power increment*

The essence of grid loss analysis based on the complex power increment is tracing the branch loss according to the source of current, then the complex proportion function of S_{loss} and S_{DGi} can be obtained. Because the complex coefficient is composed of the impedance parameters of the grid, the incremental relation between the grid loss and the power supply is obtained by the derivation of the formula. The concrete deduction process is as follows:

$$S_{loss,i} = (U_{i1} - U_{i2}) * [(-Y_{i1,i2}) * (U_{i1} - U_{i2})]^* \tag{10}$$

The load under the steady state flow is equivalent to the ground impedance, a new admittance matrix is formed by adding the earth impedance to the node admittance matrix, expressed by *YN*. The nodal voltage equation $I_s = YN * U \Rightarrow U = YN^{-1} * I_s$ is substituted into the formula (10) to obtain the formula:

$$S_{loss,i} = (U_{i1} - U_{i2}) * [(-Y_{i1,i2}) * \sum_{j=0}^{q} (YN^{-1}_{i1,j} - YN^{-1}_{i2,j}) * I_{js}]^* \tag{11}$$

Divide the formula (11) by the distributed generation output $S_{DGi} = U_i * \overset{*}{I}_{is}$, and then get the function:

$$S_{loss,i} = \sum_{j=0}^{q} \frac{U_{i1} - U_{i2}}{U_j} * [(-Y_{i1,i2}) * (YN^{-1}_{i1,j} - YN^{-1}_{i2,j})]^* * S_j \tag{12}$$

In the above formula, S_j for all power generation in the grid, including system injection and distributed generation, to simplify the representation, use G_{ij} to express the coefficients in front of S_j, subscript i is the serial number of the branch, and j is the serial number of the power after simplification, the function is:

$$S_{loss,i} = \sum_{j=0}^{q} G_{ij} * S_j \tag{13}$$

It can be seen from the formula G_{ij} that the complex proportionality factor only depends on the impedance parameter of the grid under the steady state power flow, and can be regarded as a fixed value. Therefore, the grid loss description based on the complex power increment is more accurate than the extended sensitivity method.

Summary of all network branch loss form the whole loss:

$$S_{loss} = \sum_{i=1}^{m} S_{loss,i} = \sum_{i=1}^{m} \sum_{j=0}^{q} G_{ij} * S_j = \sum_{j=0}^{q} G_{loss,j} * S_j \tag{14}$$

Spread the above formula according to the real part and the imaginary part, derivation of active power and reactive power of distributed power generation gives us the grid loss increment based on complex power increment.

2.3. An example of increment analysis of grid loss

In order to demonstrate the accuracy of the grid loss increment obtained by two methods, the 14-node distribution grid model with distributed generation is selected to perform analysis. The number of nodes in the model is n=14, and the node of system injection is selected as slack node, its number is 14. There are 4 nodes that are distributed generation access points, and their numbers are 1, 6, 7 and 9.The grid branch number is m=13. In order to maintain the safe and stable operation of the power grid, the amount of distributed generation output will not exceed 15% of the total power of the grid. It is assumed that the output of the distributed generation is 50% active power and 40% reactive power. According to the formulas of the last two sections, calculate the incremental derivative of grid loss, then, calculate the incremental value of the grid loss when the output of distributed generation is increased by 1%.For the sake of simplicity, only the derivative of the active power of the grid loss to the active power of the distributed generation is listed in the following table 1:

Table 1. Three methods comparing.

Increment value	Extended sensitivity method	Complex power increment method	Numerical method
dP_{loss}/dP_1	-0.003548856696838	-0.018556638568433	-0.030976916460812
dP_{loss}/dP_6	-0.001335291930379	-0.033946516255844	-0.059520110509137
dP_{loss}/dP_7	0.051617733291405	-0.030641780835950	-0.053696161612660
dP_{loss}/dP_9	0.002129094978441	-0.019949855624131	-0.032213682314985

After comparing the data, it can be found that the derivatives obtained by applying the complex power increment method are closer to the numerical results than those obtained by the extended sensitivity method. This shows that the method of complex power increment can make the optimization process more rapid and accurate in the optimization of power output.

3. Interior point optimization method

The complex power increment method described in the previous section is a new method for obtaining the direction of steady state power flow, this method is more accurate and faster than the extended sensitivity method, but it still needs to be combined with specific optimization methods when solving specific problems. The interior point method is a superior and widely used continuous optimization method for solving the optimization problem of distributed generation [13]. In the optimization process, the interior point method combined with the complex power increment does not need to consider the equality constraint. Due to the analysis of complex power increment is based on steady-state power flow, the power flow balance is always kept in the optimization iteration process. Thus, compared to the interior point optimal power flow, which considered the power flow balance constraints, will be simpler and faster. Compared with the heuristic algorithm, the defect of the classical mathematical optimization method is easy to fall into the local optimal solution and make the result of optimization inaccurate [14]. But considering that the capacity of the distributed generation will not exceed 15% of the total power generation capacity, the feasible region of the function variable is smaller and the possibility of falling into the local optimal solution is greatly reduced. Therefore, the optimization scheme based on the complex power increment is simpler and faster than the existing optimization methods.

In this paper, the objective function takes the profit of distributed generation as the main factor, meanwhile, considers the profit of the grid and system power supply. Currently, distributed generation owners are often not grid companies, therefore, the maximum profit of power supply is the primary purpose of its operation considerations. But the distributed power supply will affect the stability and profitability of the whole power grid, and its influence will increase with the capacity of the network increases. After the distributed power supply access to the grid, in order to guarantee the operation of them has a positive impact on the profit of power grid, when planning the output of the distributed power supply, the profit of the grid should be taken into account. Similarly, the system power is responsible for most of the active power load and almost the whole reactive power load. The increase of output of distributed generation access to the network will reduce the power supply of the system source and reduce the power factor, thus reduces the profitability of the traditional power plants. Therefore, in the optimization process, the impact of the profit on the system power (slack node) should also be taken into account. To sum up, the objective function, that is, the comprehensive profit per unit time (1hour) under steady flow, is:

$$obj : F = f_1 + k * f_2 + \lambda * f_3 \tag{15}$$

$$f_1 = \sum_{i=1}^{q} (D_i - C_i) * P_i \tag{16}$$

$$f_2 = \sum_{i=1}^{l} D_{Li} * P_{Li} - \sum_{i=1}^{q} D_{bi} * P_i - D_{b0} * P_0 \tag{17}$$

$$f_3 = D_0 * P_0 - (a_2 * P_0^2 + a_1 * P_0 + a_0) \tag{18}$$

In the objective function, D_i is the power on line price, that is, the power supplier selling price. In this paper, the price of wind power is 0.52￥/kWh, and the photoelectric power price is 1￥/kWh. D_{bi} is the purchase price of Power Grid Corp, due to policy factors, distributed power generation has central and local financial subsidies, therefore, in addition to the system power supply, there is a big gap between D_i and D_{bi}. Taking D_{bi} as 0.3￥/kWh. D_{Li} is the selling price of Power Grid Corp, which takes 0.6￥/kWh, and C_i is the comprehensive cost of distributed generation, according to the data in the literature [15], the cost of photovoltaic power generation is 0.2￥/kWh, and the cost of wind power generation is 0.15￥/kWh. a_1, a_2 and a_3 are the cost coefficient of the system power generation. To simplify calculation, they are 0.008, 0.08 and 0, k and λ are the effects of grid profit and system power profit on the objective function, taking 0.5 and 0.25. In the function, P_i is the active power output of the power supply, the unit is kW, l is the number of the load, and q is the number of the distributed power supply.

Before to optimize the objective function, considering the price of different nodes are not the same, in function f_2, the cost calculation node and the charging calculation node are different, therefore, the cost should be recalculated according to the conservation principle of fee flow [16]. The cost is expressed by C , and the active power is expressed by P, the conservation formula of nodes and branches are:

$$\sum C_i^{in} * P_i^{in} = \sum C_i^{out} * P_i^{out} \tag{19}$$

$$C_l^{in} * P_l^{in} = C_l^{out} * P_l^{out} \tag{20}$$

For fairness, the active power flow from the same node is assumed the same price, the function f_2 can be written in the following form:

$$f_2 = \sum_{i=1}^{l} \sum_{j=0}^{q} (D_{L,i} - D_{Lb,ij}) * P_{L,ij} \tag{21}$$

In this way, the variables in the object function are independent variables except P_0, then replace it by the formula (1).

For the constraints of the function, we consider the upper and lower limits of the distributed power supply, the upper and lower limits of the network node voltage, and the upper and lower limits of the system power supply. Due to the distributed power supply capacity is small, the line power limit is not considered in this paper, considering that the power factor of the distributed power supply should not be too small, the reactive power limit shall not exceed 20% of the active power limit when the power limit is setted. The specific restrictions are as follows:

$$
\begin{aligned}
P_{i,\min} &\le P_i \le P_{i,\max} \\
Q_{i,\min} &\le Q_i \le Q_{i,\max} \\
U_{i,\min} &\le U_i \le U_{i,\max} \\
P_{0\min} &\le P_0 \le P_{0\max}
\end{aligned}
\tag{22}
$$

When the constraints are taken into consideration, the initial values of the variables of the function are set within the feasible region, then the optimization results in the current situation can be obtained after iteration.

4. Example analysis

In order to demonstrate the analysis process of power output optimization, the distribution network model adopted in this paper is the model used in the third section, second chapter. Nodes 7 and 9 are wind turbines, nodes 1 and 6 are photovoltaic generators, their output limit is: active power 1MW and reactive power 0.2MVar. The system generation output limit is: active power 60MW and reactive power 12MVar, the per-unit value of voltage limit of each node is 0.95~1.05. The extended sensitivity algorithm and the complex power increment algorithm are used respectively to optimize the results:

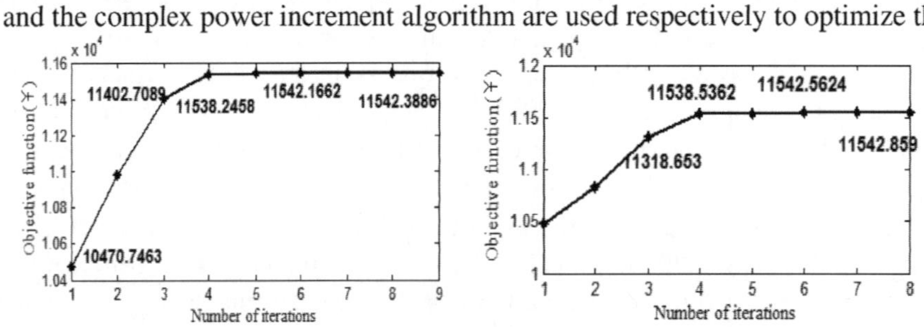

Figure 1. Comparison of optimization results.

Figure 1 shows: the diagram on the left is the result of the extended sensitivity method, the right one is the result of the complex power increment method, after comparison, it is found that the complex power increment optimization results are more accurate and more efficient. For further explanation, the direction of flow adjustment obtained by the method of complex power increment is more accurate, in the adjustment process, if we observe the power supply output adjustment of the two methods, we will find that the adjustment process of the active power is approximately the same, but there are differences in the adjustment of reactive power, and even the opposite direction of adjustment. The difference of reactive power adjustment will directly affect the overall grid loss, in the process of adjustment, the grid loss of the two methods are compared, listed in the figure 2:

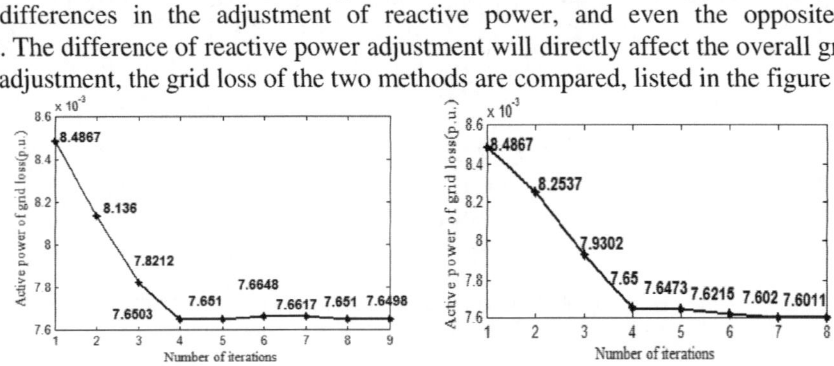

Figure 2. Comparison of grid loss active power adjustment.

In the optimization process, if we apply the method of complex power increment, the grid loss will decrease, if we apply the extended sensitivity method, the network loss will fluctuates. The comparison of the three sets of graphs shows that the optimization method based on the complex power increment is superior to the optimization method based on the extended sensitivity in the optimization speed and the accuracy of results.

5. Conclusion

First of all, the deficiency of sensitivity algorithm is pointed out in this paper, a new method based on complex power increment is proposed, which can obtain the derivative of variables more accurately.

Secondly, a power relation equation is established under steady-state power flow, according to the nodal voltage equation, we can replace and decompose the equation, then form a complex scaling function between the power supplies. This complex coefficient is composed of distribution network impedance, and can be regarded as fixed value in steady state power flow. Therefore, the power relation in steady state flow can be represented by linear equation.

Last, based on the idea of complex power increment, the coupling relationship between the parameters of the nodes is taken into account, therefore, the results are more accurate. The linearity of the equation makes the process more rapid, and this is also the reason why the method is superior to the extended sensitivity algorithm.

References

[1] Information on http://www.nea.gov.cn/2016-11/07/c_135811323.htm， 2016-11-07
[2] Li Ya lou, Zhou Xiao xin, Lin Ji ming, Jiang Yi guo and Sun De dong, A Review of New Energy Power Generation Part in 2008 IEEE PES General Meeting, Power System Technology. 32 (20) 1-7.
[3] Wen Sheng, Optimal Deployment and Dispatch of Distributed Generation, Shanghai Jiao Tong University, Shanghai, 2012.
[4] Sultan S. Al Kaabi, H. H. Zeineldin, Vinod Khadkikar. Planning Active Distribution Networks Considering Multi-DG Configurations, IEEE Transactions on Power Systems. 29(2): 785-793.
[5] Wu Jia sheng, Study on Modeling and Appliance of Optimal Dispatch for Distribution Network with Distributed Generation, North China Electric Power University, Beijing, 2014.
[6] Liu Yuan, Optimal Scheduling for Distributed Energy Resource, Hunan University, Hunan, 2013.
[7] Li Ran, Ma Hui zhuo, Zhu Jin yao, Wang Yi he, Hu Bo, Multi-Objective Optimization for DG Integration into Distribution System, Electric Power Automation Equipment. 34 (1) 1-7.
[8] Liu Zi fa, Liu Gang, Liu Xing, Coordinated Optimal Dispatching of Distributed Generation Based on Quantum Differential Evolution Algorithm, Power System Technology. 37 (7) 1922-1928.
[9] Wan Wei, Huang Da wei, The Coordination Optimization of Distribution Networks with Dispatched Distributed Generators, Transactions of China Electrotechnical Society. 30 (12) 429-433.
[10] Zheng Zhang hua, Ai Qian, Gu Cheng hong, Jiang Chuan wen, Multi-objective Allocation of Distributed Generation Considering Environmental Factor, Proceedings of the CSEE. 29 (13) 23-28.
[11] Wei Guo hua, Network Losses Analysis Based on Sensitivity and Network Parameter Comprehensive Optimization, Yanshan University, Hebei, 2010.
[12] M.Braun. Reactive Power Supply by Distributed Generators. 2008 IEEE Power and Energy Society General Meeting, Pittsburgh, USA, 2008, 1-8.
[13] Wang Xi fan, Modern power system analysis, first ed., Sciencep, Beijing, 2003.

[14] Wang Shou xiang, Wang Hui, Cai Sheng xia, A Review of Optimization Allocation of Distributed Generations Embedded in Power Grid, Automation of Electric Power Systems. 33 (18) 110-113.
[15] Qian Ke jun, Yuan Yue, Shi Xiao dan, Zhou Cheng ke, Ju Ping, Environmental Benefits Analysis of Distributed Generation, Proceedings of the CSEE. 28 (29) 11-15.
[16] Sun Hui jiao, Bao Hai, Wang Xiao jun, The Influence of Network Equivalence on Power Transmission Cost, Proceedings of the CSEE. 27 (4) 79-84.

A Calculation Method of Electric Distance and Subarea Division Application Based on Transmission Impedance

G J Fang[1] and H Bao[1]

[1]School of Electrical & Electronic Engineering, North China Electric Power University, Beijing 102206, China

18001059951@163.com; baohai@ncepu.edu.cn

Abstract. The widely used method of calculating electric distances is sensitivity method. The sensitivity matrix is the result of linearization and based on the hypothesis that the active power and reactive power are decoupled, so it is inaccurate. In addition, it calculates the ratio of two partial derivatives as the relationship of two dependent variables, so there is no physical meaning. This paper presents a new method for calculating electrical distance, namely transmission impedance method. It forms power supply paths based on power flow tracing, then establishes generalized branches to calculate transmission impedances. In this paper, the target of power flow tracing is S instead of Q. Q itself has no direction and the grid delivers complex power so that S contains more electrical information than Q. By describing the power transmission relationship of the branch and drawing block diagrams in both forward and reverse directions, it can be found that the numerators of feedback parts of two block diagrams are all the transmission impedances. To ensure the distance is scalar, the absolute value of transmission impedance is defined as electrical distance. Dividing network according to the electric distances and comparing with the results of sensitivity method, it proves that the transmission impedance method can adapt to the dynamic change of system better and reach a reasonable subarea division scheme.

1. Introduction

In recent years, many domestic and foreign power systems have appeared voltage instability or voltage collapse causing the local power outages, so the voltage stability issues have attracted more and more attention [1-3]. Voltage classification control is an effective method of reactive power/voltage control. The premise of voltage classification control is subarea division, which is based on the electrical distance [4]. In 1989, P. Lagonotte presented the concept and formula of electric distance based on sensitivity and applied it to subarea division for the first time [5]. Now most of common methods are based on the idea of P. Lagonotte, but only improve in some ways for accuracy and rationality [6-8].

Sensitivity method can obtain the relation of the dependent variable to the independent variable by calculating derivatives [9]. The sensitivity matrix is the result of linearization and based on the hypothesis that the active power and reactive power are decoupled, so it is inaccurate. The ratio of two partial derivatives is not the relationship between two dependent variables, so there is no physical meaning. In addition, most papers do not form a unified standard for calculating sensitivity indexes and lack of mathematical proof.

Therefore, this paper presents a new method of calculating electric distances called transmission impedance method. It forms power supply paths based on power flow tracing, then establishes generalized branches to calculate transmission impedances. The electrical distances are calculated by sensitivity method and transmission impedance method, and are applied to subarea division of IEEE three machines nine nodes system. Through the comparison and analysis of two results, the accuracy and feasibility of the new method can be seen.

2. Sensitivity method

The most widely used method of calculating electrical distance is the sensitivity method. The calculation process is as follows:

$$\begin{bmatrix} \Delta P \\ \Delta Q \end{bmatrix} = \begin{bmatrix} H & N \\ J & L \end{bmatrix} \begin{bmatrix} \Delta\theta \\ \Delta U / U \end{bmatrix} \tag{1}$$

Deformed the above matrix equation as follows:

$$\begin{bmatrix} \Delta\theta \\ \Delta U / U \end{bmatrix} = \begin{bmatrix} A & B \\ C & S \end{bmatrix} \begin{bmatrix} \Delta P \\ \Delta Q \end{bmatrix} \tag{2}$$

There into, S is called reactive power/voltage sensitivity matrix, and its solution is as follows:

$$S = \left[\frac{1}{U} \frac{\partial U}{\partial Q} \right] = \left[L - JH^{-1}N \right]^{-1} \tag{3}$$

According to the reactive/voltage sensitivity, the electrical distance is defined as follows:

$$\Delta U_i = \alpha_{ij} \cdot \Delta U_j \tag{4}$$

$$\alpha_{ij} = \left(\frac{\partial U_i}{\partial Q_j} \right) / \left(\frac{\partial U_j}{\partial Q_j} \right) \tag{5}$$

In order to ensure the positive definiteness and symmetry of the electric distance matrix, the following mathematical treatment is carried out:

$$D_{ij} = -\log\left(\alpha_{ij}\right) \tag{6}$$

$$D_{ij} = D_{ji} = -\log\left(\alpha_{ij} \cdot \alpha_{ji}\right) \tag{7}$$

From above we can see that there are several problems:

(1) In the process of linearization of the power flow equation, the accuracy is affected caused by omitting the higher order term.

(2) There is no physical meaning for the ratio of two partial derivatives. $\frac{\partial U_i}{\partial Q_j}$ represents that changes of Q_j cause changes of U_i; $\frac{\partial U_j}{\partial Q_j}$ represents that changes of Q_j cause changes of U_j; $\frac{\partial U_i}{\partial U_j}$ represents that changes of U_j cause changes of U_i. The sensitivity method considers $\left(\frac{\partial U_i}{\partial Q_j} \right) / \left(\frac{\partial U_j}{\partial Q_j} \right) = \frac{\partial U_i}{\partial U_j}$ correct. However, in fact, because the independent variable is eliminated, the calculation result is only a ratio, which cannot explain the causal relationship between U_i and U_j.

(3) For the same transmission process, the electric distance of forward direction should be equal to the one of inverse direction. The sensitivity method lacks proof only by mathematical processing to ensure symmetry.

3. Transmission impedance method

3.1. Generalized branch and transmission impedance

The generalized branch is relative to the narrow branch, of which two nodes are not necessarily directly connected [10]. As we all know, for a branch, the injected current is equal to the output current. So, if two nodes are not directly connected but meet the condition, we may can describe the electrical relationship between the two nodes from a generalized perspective. As shown in Figure 1:

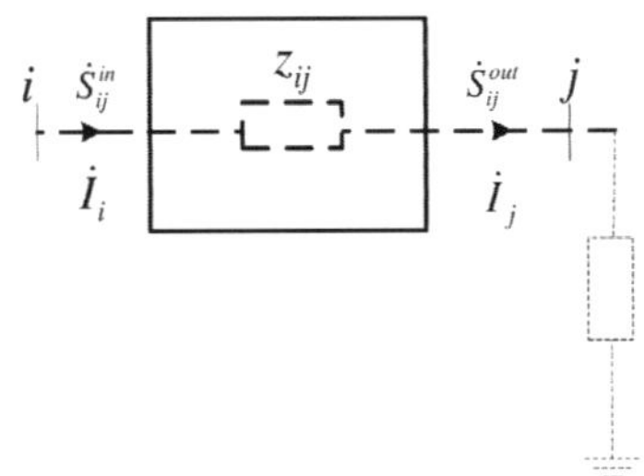

Figure 1. Generalized branch

If $\dot{I}_i = \dot{I}_j$, then there is a virtual branch between node i and node j, which is called generalized branch. The impedance of the generalized branch can characterize the voltage loss caused by the current flowing from node i to node j, which is called transmission impedance.

3.2. Power flow tracing

Power flow tracing is an effective method to establish generalized branches and obtain transmission impedances. Based on the power flow results, firstly form multiple single-power supply networks by downstream tracking. Then, for each single-power supply network, do upstream tracking to find the power supply paths from each power source to each load. Combining with voltages, we can calculate the transmission impedances between every two nodes on a power supply path [11-13]. The transmission impedance between two nodes in original network is equal to the parallel result of transmission impedances between the two nodes of all power supply paths. Obviously, the transmission impedance is a finite value only when there is power transmission between nodes. If there is no transmission, the impedance is infinite.

In this paper, the target of power flow tracing is $\dot{S}$ instead of Q, partly because Q itself has no direction. In addition, the grid delivers complex power so that $\dot{S}$ contains more electrical information than Q.

3.3. The bidirectional description of power transmission and the definition of electrical distance

3.3.1. Forward power transmission

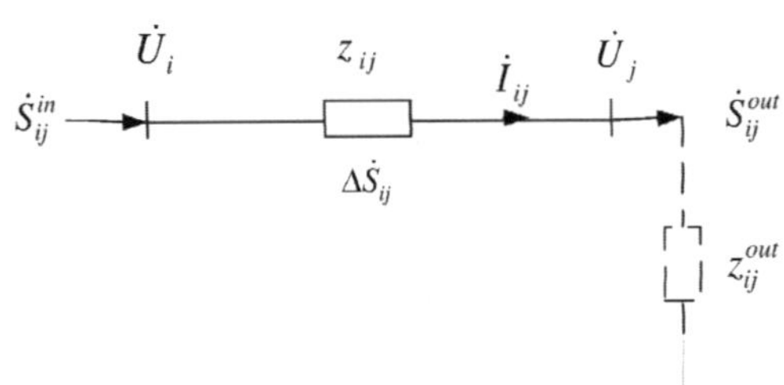

Figure 2. Forward power transmission

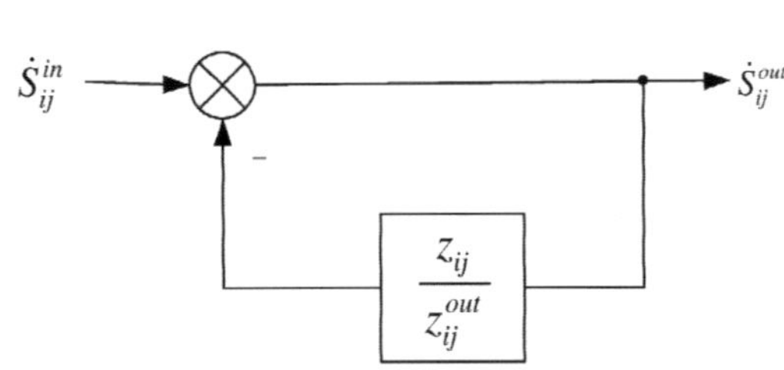

Figure 3. Block diagram of forward transmission

According to Figure 2, there are

$$\dot{S}_{ij}^{in} = \Delta\dot{S}_{ij} + \dot{S}_{ij}^{out} \tag{8}$$

$$\dot{S}_{ij}^{out} = \dot{U}_j \cdot (\dot{I}_{ij})^* = z_{ij}^{out} \cdot \dot{I}_{ij} \cdot (\dot{I}_{ij})^* \tag{9}$$

$$\Delta\dot{S}_{ij} = \left(\dot{U}_i - \dot{U}_j\right)\cdot\left(\dot{I}_{ij}\right)^* = z_{ij}\cdot \dot{I}_{ij}\cdot\left(\dot{I}_{ij}\right)^* \tag{10}$$

According to formulas(9)(10), there is

$$\Delta\dot{S}_{ij} = \frac{z_{ij}}{z_{ij}^{out}} \cdot \dot{S}_{ij}^{out} \tag{11}$$

According to formulas(8)(11), there is

$$\dot{S}_{ij}^{in} = \left(1 + \frac{z_{ij}}{z_{ij}^{out}}\right) \cdot \dot{S}_{ij}^{out} \tag{12}$$

Deform the formula(12), there is

$$\frac{\dot{S}_{ij}^{out}}{\dot{S}_{ij}^{in}} = \frac{1}{1 + \dfrac{z_{ij}}{z_{ij}^{out}}} \tag{13}$$

It can be seen that the formula(13) describes the relationship between the input power and the output power of branch $i-j$. Draw the block diagram as Figure 3. We can see that the block diagram contains a negative feedback part that represents the ratio of the loss power to the output power. The numerator of feedback part is the transmission impedance of branch $i-j$.

3.3.2. Reverse power transmission

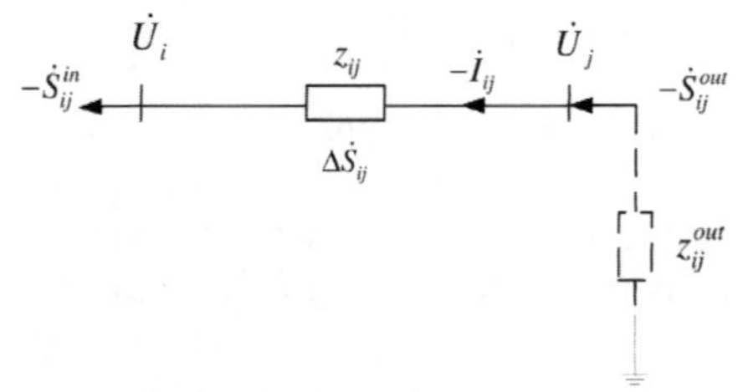

Figure 4. Reverse power transmission

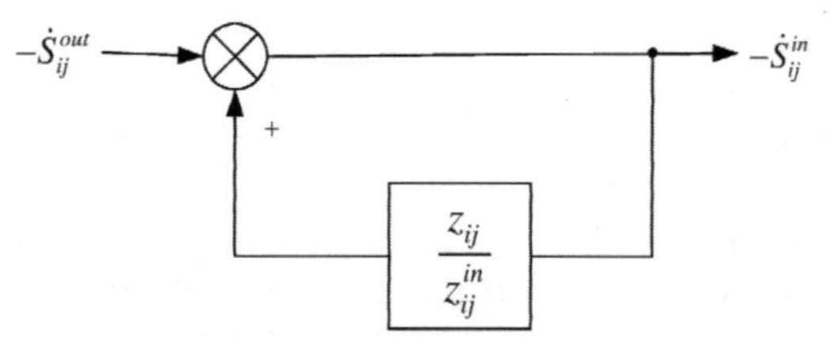

Figure 5. Block diagram of reverse transmission

According to Figure 4, imitate the derivation process of formulas(8)~(13) and we can get

$$\frac{-\dot{S}_{ij}^{in}}{-\dot{S}_{ij}^{out}} = \frac{1}{1 - \dfrac{z_{ij}}{z_{ij}^{in}}} \tag{14}$$

Similarly, it can be seen that the formula(14) describes the relationship between the input power and the output power of branch $j-i$. Draw the block diagram as Figure 5. We can see that the block diagram contains a positive feedback part that represents the ratio of the loss power to the input power. The numerator of feedback part is the transmission impedance of branch $j-i$.

It can be found that the numerators of feedback parts of two block diagrams are all the transmission impedances. Whereas distances are generally defined as scalars for comparison and applied to partitions, we take the absolute value of transmission impedance as electric distance.

4. Case study

In order to verify the feasibility of the proposed method, the IEEE three machines nine nodes system is taken as an example to apply this method and achieve reactive power partitioning. The shortest distance clustering method is adopted for the partition method [14].

4.1. Calculation results and subarea application

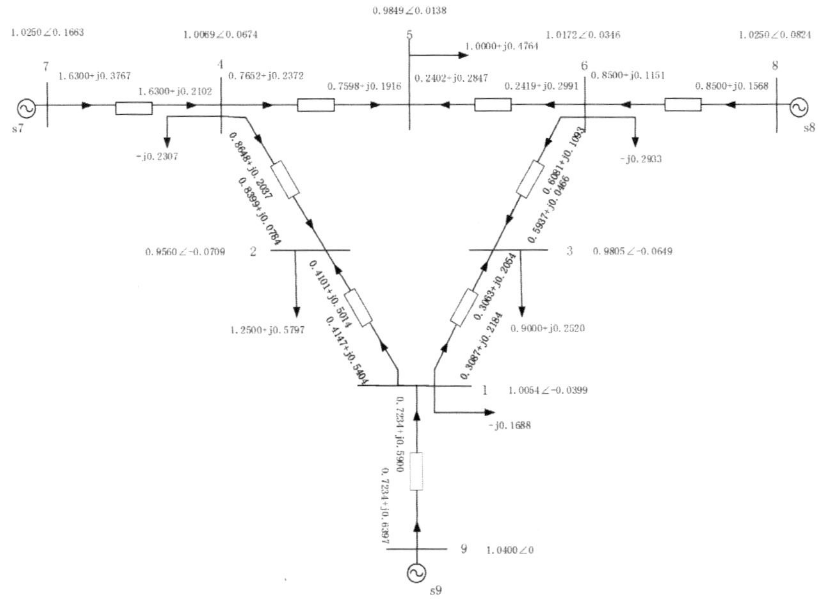

Figure 6. IEEE three machines nine nodes system

By downstream tracking and upstream tracking, we can find power transmission paths from each power source to each load and calculate the transmission impedances. Shown as follows:

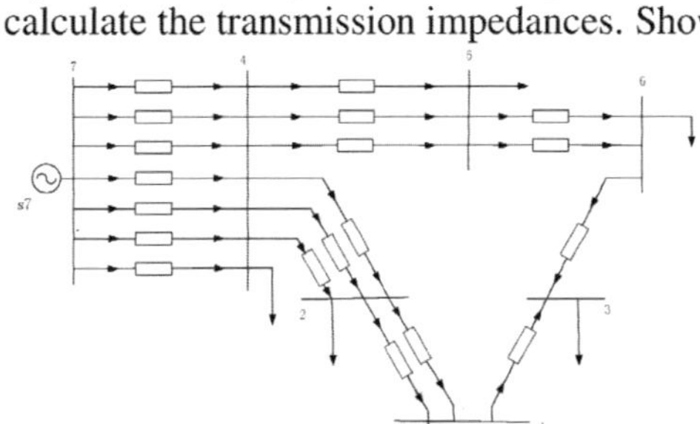

Figure 7. Transmission impedances by s 7

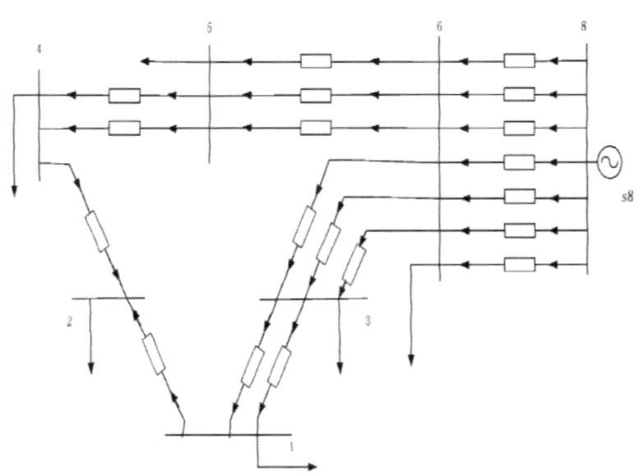

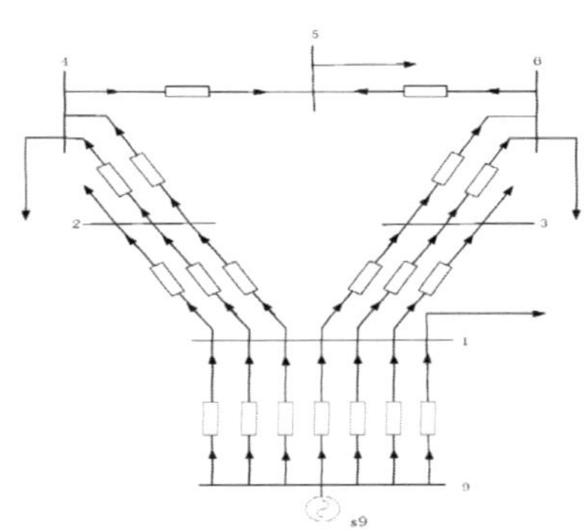

Figure 8. Transmission impedances by s **8** **Figure 9.** Transmission impedances by s 9

As shown figures (Figure 6, Figure 7, Figure 8 and Figure 9) above, we can get transmission impedances between nodes under the influence of each power source. For the original network, the transmission impedance between two nodes is equal to the parallel result of transmission impedances between the two nodes under the influence of 3 power sources. We take the absolute value of

IOP Conf. Series: Earth and Environmental Science **104** (2017) 012006 doi:10.1088/1755-1315/104/1/012006

transmission impedance as electric distance. In order to compare with the sensitivity method, the electrical distances between the PQ nodes are only listed in Table 1 and Table 2.

Table 1. The electrical distances of transmission impedance method

Electrical distance	node1	node2	node2	node3	node4	node5	node6
node1	0	0.0856	0.0936	0.6951	0.1886	1.3268	0
node2	0.0856	0	0.3294	0.1641	0.7306	0.3303	0.0856
node3	0.0936	0.3294	0	0.2908	0.3010	0.1744	0.0936
node4	0.6951	0.1641	0.2908	0	0.0725	0.2538	0.6951
node5	0.1886	0.7306	0.3010	0.0725	0	0.1015	0.1886
node6	1.3268	0.3303	0.1744	0.2538	0.1015	0	1.3268

Table 2. The electrical distances of sensitivity method

Electrical distance	node1	node2	node2	node3	node4	node5	node6
node1	0	1.2070	1.2873	2.9487	3.4520	3.0799	0
node2	1.2070	0	2.3910	1.9812	2.9118	3.4616	1.2070
node3	1.2873	2.3910	0	3.4604	3.2376	2.0962	1.2873
node4	2.9487	1.9812	3.4604	0	1.1508	2.5794	2.9487
node5	3.4520	2.9118	3.2376	1.1508	0	1.5639	3.4520
node6	3.0799	3.4616	2.0962	2.5794	1.5639	0	3.0799

Cluster according to the shortest distance and here are the dendrograms:

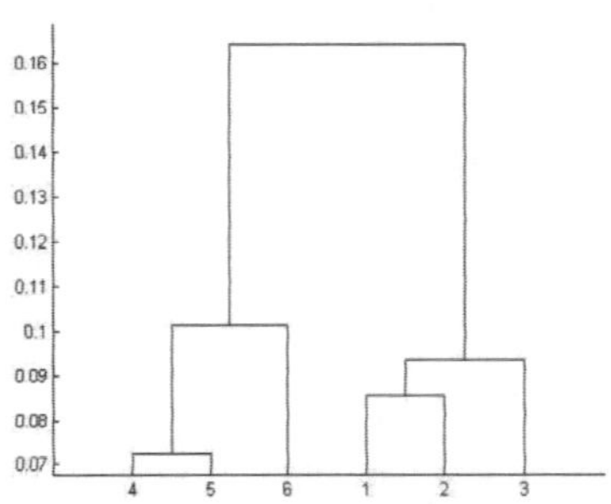

Figure 10. Dendrogram of new method

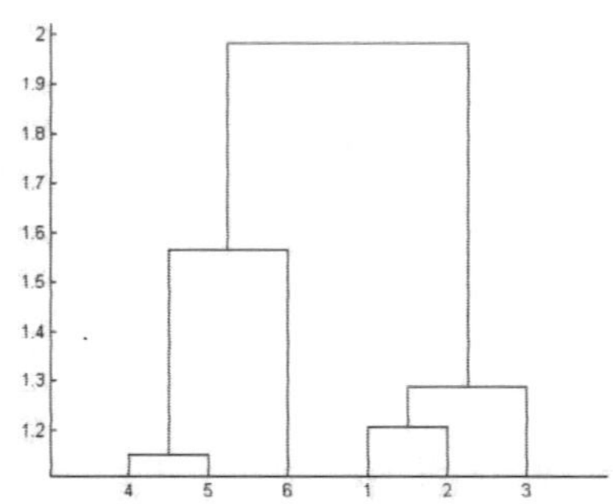

Figure 11. Dendrogram of sensitivity method

Comparing Figure 10 with Figure 11, it can be seen that the clustering processes of transmission impedance method and sensitivity method are consistent, so the partition results are always same no matter the system is divided into two or three areas, which fully demonstrates the feasibility of transmission impedance method applied for electrical distance calculation and subarea division. According to the merging principle of PV nodes [15], the generators of 7, 8 and 9 are divided into different regions, and the final partition results are shown in Figures 12 and 13:

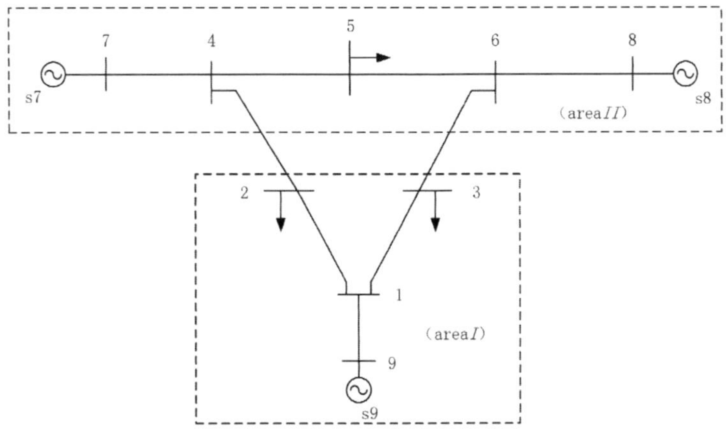

Figure 12. partition results of 2

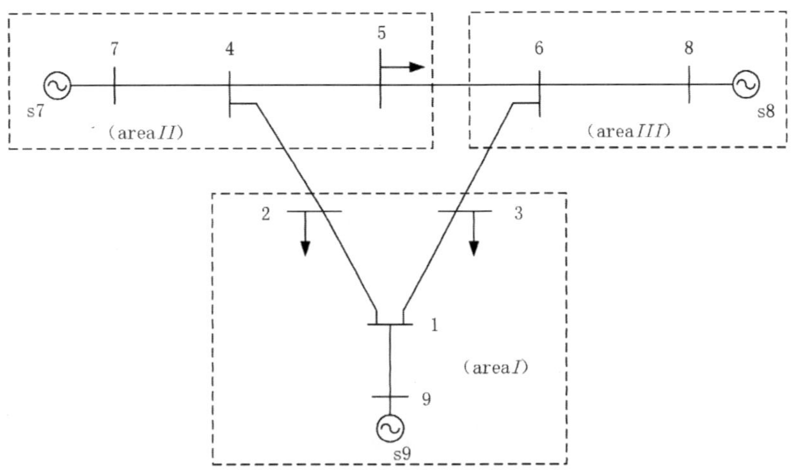

Figure 13. partition results of 3

4.2. Results after adjusting power flow

Increase the output of generator 7, reduce the output of generator 8, and adjust the loads properly. The results of electrical distance calculated by transmission impedance method and sensitivity method are listed in Table 3 and Table 4.

Table 3. The electrical distances of transmission impedance method after adjusting power flow

Electrical distance	node1	node2	node2	node3	node4	node5	node6
node1	0	0.0856	0.0936	0.5744	0.1591	0.9293	0
node2	0.0856	0	0.2411	0.1641	0.5182	0.3153	0.0856
node3	0.0936	0.2411	0	0.2823	0.2956	0.1744	0.0936
node4	0.5744	0.1641	0.2823	0	0.0725	0.2765	0.5744
node5	0.1591	0.5182	0.2956	0.0725	0	0.1015	0.1591
node6	0.9293	0.3153	0.1744	0.2765	0.1015	0	0.9293

Table 4. The electrical distances of sensitivity method after adjusting power flow

Electrical distance	node1	node2	node2	node3	node4	node5	node6
node1	0	1.2285	1.3013	3.1047	3.4780	2.9864	0

node2	1.2285	0	2.4252	3.4874	2.9275	2.0034	1.2285
node3	1.3013	2.4252	0	2.1091	3.2513	3.4902	1.3013
node4	3.1047	3.4874	2.1091	0	1.5539	2.5672	3.1047
node5	3.4780	2.9275	3.2513	1.5539	0	1.1405	3.4780
node6	2.9864	2.0034	3.4902	2.5672	1.1405	0	2.9864

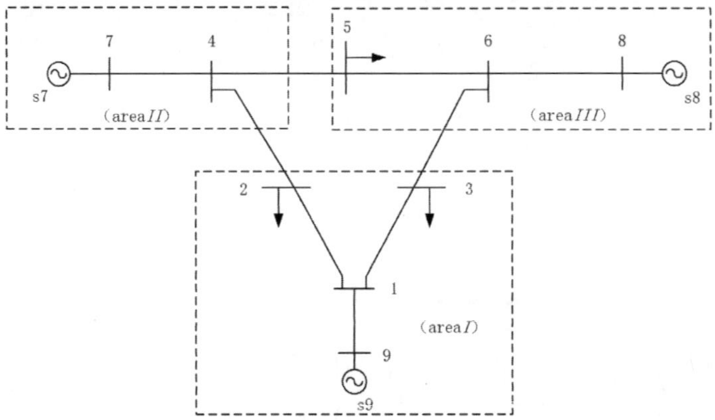

Figure 14. 3 partitions of sensitivity method after adjusting power flow

After clustering according to the shortest distance, it can be found that after the power flow adjustment, when the partition number is 2, the partitioning results of two methods all remain unchanged. When the partition number is 3, the partitioning result of the transmission impedance method is unchanged, while the result of sensitivity method is different where node 5 belongs to area *III*, as shown in Figure 14. Although the partitioning results are not identical before and after the adjustment, the partition schemes all satisfy the evaluation of the partition results [16].

Because the partition results of two methods are different for node 5, now we can identify which partition scheme is more accurate according to reactive power compensation ability to node 5 of generators in area *II* and *III* [17]. Increasing the reactive power of load 5 by 10%, the reactive power changes of generator 7 and 8 are obtained: $\Delta Q_{G7} = 0.001$， $\Delta Q_{G3} = 0.0004$. It can be seen that generator 7 has more reactive power compensation to load 5, so the electrical coupling relationship between them is stronger, and node 5 should be divided into area *II*. It proves that the subarea division based on the transmission impedance method can adapt to the dynamic change of the system better.

5. Conclusion

The widely used sensitivity method of calculating electric distance is inaccuracy and lack of physical meaning. This paper presents a new method for calculating electrical distance, namely transmission impedance method. It forms power supply paths based on power flow tracing, then establishes generalized branches to calculate transmission impedances. By describing the power transmission relationship of the branch and drawing block diagrams in both forward and reverse directions, it can be found that the numerators of feedback parts of two block diagrams are all the transmission impedances. To ensure the distance is scalar, the absolute value of transmission impedance is defined as electrical distance. Dividing network according to the electric distance and comparing with the results of sensitivity method, it proves that the new method can adapt to the dynamic change of system better and reach a reasonable subarea division scheme.

References

[1] Mao A J, Zhang G L, and Lv Y C 2012 Analysis on large-scale blackout occurred in south America and north Mexico interconnected power grid on Sept. 8, 2011 and lessons for electric power dispatching in China *Power System Technology* **36** 74-78

[2] Lin W F, Tang Y, and Sun H D 2011 Blackout in Brazil power grid on February 4, 2011 and inspirations for stable operation of power grid *Automation of Electric Power Systems* **35** 1-5

[3] Tang Y, Bu G Q, and Yi J 2012 Analysis and lessons of the blackout in Indian power grid on July 30 and 31, 2012 *Proceedings of the CSEE* **32** 167-174+23
[4] Bo Q 2006 The research on area partition of voltage hierarchical *North China Electric Power University, Beijing*
[5] P. Lagonotte 1989 Structural analysis of the electrical system: application to secondary voltage control in France *IEEE Transactions on Power Systems* **4** 479-485
[6] Hang L, Anjan B, and Vaithianathan V 2000 A fast voltage security assessment method using adaptive bounding *IEEE Transactions on Power Systems* **15** 1137-1141
[7] Ruan J Q, Wang J X 2008 Reactive power planning based on power grid partition and comprehensive sensitivity *Yunnan Electric Power* **6** 5-7
[8] Yao Z X, Tu H Y, and Xu G Y. 1997 Sensitivity based optimal reactive power flow *Automation of Electric Power Systems* **11** 19-21
[9] Miao F X, Guo Z Z 2007 A survey of sensitivity technique and its application in power systems analysis and control *RELAY* **35** 72-76
[10] Bao S X 1987 Generalized feedback theory *HIET Journal* **2** 129-137
[11] Zhou C 2011 A power flow tracing method based on circuit analysis of power supply path *North China Electric Power University, Beijing*
[12] Yan L, Bao H 2011 Algorithm of power distribution factor based on current distribution *Proceedings of the CSEE* **1** 80-85
[13] Yan L 2010 Algorithm of power distribution factors based on current distribution *North China Electric Power University, Beijing*
[14] Bao W, Zhu T, and Zhao C 2016 A three-stage network partition method for secondary voltage control based on agglomerative analysis *Automation of Electric Power Systems* **5** 127-132
[15] Liu X H 2015 Research on distribution network reactive optimization based on network partitioning *Southwest Jiaotong University*
[16] Zhao J Q, Liu F C, and Deng Y 2010 Network partitioning for reactive power voltage control based on a mapping division algorithm *Automation of Electric Power Systems* **34** 36-39
[17] Li L, Huang Y Q, and Dong J D 2010 Optimization method of power network partitioning based on voltage/var control *Power System Protection and Control* **38** 88-91

An Equivalent Fracture Modeling Method

Li Shaohua[1], Zhang Shujuan[2], Yu Gaoming[3] and Xu Aiyun[2]

[1]Faculty of Earth Science, Yangtze University, wuhan, 430100, China
[2]Research Center of Huabei Field, renqiu, 062552, China
[3]Faculty of Petroleum Engineering, Yangtze University, wuhan, 430100, China

*Corresponding e-mail: jpishli@sina.com

Abstract. 3D fracture network model is built based on discrete fracture surfaces, which are simulated based on fracture length, dip, aperture, height and so on. The interesting area of Wumishan Formation of Renqiu buried hill reservoir is about 57 square kilometer and the thickness of target strata is more than 2000 meters. In addition with great fracture density, the fracture simulation and upscaling of discrete fracture network model of Wumishan Formation are very intense computing. In order to solve this problem, a method of equivalent fracture modeling is proposed. First of all, taking the fracture interpretation data obtained from imaging logging and conventional logging as the basic data, establish the reservoir level model, and then under the constraint of reservoir level model, take fault distance analysis model as the second variable, establish fracture density model by Sequential Gaussian Simulation method. Increasing the width, height and length of fracture, at the same time decreasing its density in order to keep the similar porosity and permeability after upscaling discrete fracture network model. In this way, the fracture model of whole interesting area can be built within an accepted time.

1. Introduction

The discrete fracture network model has been get more and more applications in the research of fractured reservoir[1], this method is the current mainstream technology to describe the fractures, with the advantages of collaborative modeling through fuse data of multidisciplinary, it can integrate outcrop, core, seismic, logging, geology, drilling and production data, so that understanding fractures from different point of view. The discrete fracture network modeling method construct fracture model through simulate the fracture network which composed of fracture slices, the fractures have different shapes, sizes, width and orientation, which realized from the geometry of the fracture system to the flow behavior of real effective detailed description. The technique not only requires the reasonable description of the geometry, size and width of fractures, but also needs to predict the fracture density reasonably, which is the basis and core of fracture modeling[2-3].

The fracture system of Wumishan Formation reservoir in Renqiu buried-hill is very complicated, the key parameters such as length, height, density etc, is difficult to obtain directly through wells data. Tong Hengmao improved the accuracy of prediction results by studying fractures of similar outcrop[4]. Through field observation, the fractures are divided into 4 grades, in which the micro-fractures are usually not reflected in the fracture model because the height is too small. The spacing of the large fractures is about 10 meters, and the density of median fractures is 5-20 stripes per meter, the density of small fractures is 5-50 stripes per meter. Zhang Shujuan classify the fractures into large scale, medium scale and small scale fractures according to the fracture development level, the large and medium scale fractures are described through seismic data, the ant tracking technology is used to simulate the large and medium scale fractures[5]. Cheng Chao introduced the process of fracture modeling by "ant tracking" technology, and verified the simulation results through imaging logging data. Discrete fracture network modeling technology is usually used to simulate the small scale fractures which can not be recognized by seismic data. The small scale fractures are very developed in Wumishan Formation reservoir in Renqiu buried-hill, which have been proved by outcrop, drilling and production dynamics data[6]. The reservoir area and density of small scale fracture is large, and the strata is very thick, it will takes a lot of computing time to simulate and build the fracture network model for each fractures, the upscaling of fracture model will also very time-consuming. In view of this problem, this paper proposed the equivalent fractures modeling method, increasing the width, height and length of fracture, at the same time decreasing its density, and determine the specific parameter settings through the simulation experiment, by this method, the simulation time is greatly reduced and the porosity and permeability of equivalent model is similar with the original fine model. The method is applied to the small scale fracture modeling of Wumishan Formation reservoir in Renqiu buried-hill, greatly reduced the computation time and meets the needs of the fracture modeling in the study area.

2. The basic principle of equivalent fracture modeling

Discrete fracture network modeling method is a kind of geostatistics modeling method[7], this method produces mainly as a result of many properties such as fractures azimuth, inclination, length, density, width and other information is very difficult to uniquely determined, using deterministic modeling methods are difficult to characterize the statistical characteristics of these parameters, and thus requires to descript and modeling the fractures based on the geostatistics method. In DFN modeling, fractures density is one of the important parameters used to describe the development degree of fracture. Generally, where the fracture density is large, fractures are concentrated and high permeable zones are formed. Dense layers are often formed in where the fracture density is low.

Line density, surface density and volume density is the mainly method to descript fracture density[8]. Line density refers to the number of fractures in the unit length of the direction that perpendicular to the fracture. For example, the fracture density of small and medium scales is generally 5-50 stripes per meter in Renqiu buried-hill reservoir. The fracture network model of the whole reservoir should be established according to the requirements of the reservoir development plan of Wumishan Formation reservoir in Renqiu buried-hill. The DFN modeling technique is adopted for the small scale fractures. The area of the model is 57Km^2, thickness of strata is more than 2000 meters, at least hundreds of millions of fractures need to be simulated if it is simulate in strict accordance with the fracture density of each fractures, the general computer cannot meet the requirements, and the computation time is not acceptable. To solve this problem, this paper proposed equivalent fractures modeling method, the core idea of this method is the equivalent of multiple fractures to single fracture, its essence is to reduce the fractures density, and it will greatly reduce the time required for fracture modeling. The essence of the porosity and permeability modeling method is a kind of equivalent modeling method, it is impossible to simulate the porosity and permeability for each grain of sand, and also unlikely to direct simulate the porosity and permeability which the logging curves can explain (8 data per meter). Before the simulation, the porosity, permeability and other parameters are equivalent to the simulated grid by using the logging upscaling method, and then the grid is simulated.

The main input parameters of reservoir simulation are porosity and permeability model, rather than directly using the fracture network model. The equivalent fracture modeling method can reduce the density of fractures, and the simulated fracture porosity and permeability will also decrease. The porosity and permeability of single fracture are mainly related to the length, height, width and density of fractures. Formulas (1) and (2) are theoretical formulas for the calculation of porosity and permeability of single fracture[9],

$$\emptyset_f = \frac{e.\lambda.l}{D.V_T} \tag{1}$$

$$K_f = \frac{e_1^3}{12D_1}\cos\alpha + \frac{e_2^3}{12D_2}\cos\beta + \cdots \tag{2}$$

Φ_f is fracture porosity, K_f is fracture permeability, V_T is rock volume, λ is fracture length, l is fractures height, e is fractures width, D is fractures spacing, α and β is the angle between fluid pressure gradient axis and the fractures. From these two formulas, it can be seen that the porosity of the single set of fractures is mainly related to the length, height, width and density of fractures, and the permeability is mainly related to the fracture width and density. The equivalent fracture method proposed in this paper is actually to increase the width, height and length of fracture, at the same time decreasing its density in order to keep the similar porosity and permeability after upscaling discrete fracture network model.

3. Case study

Renqiu buried hill reservoir is located in the north of the Raoyang Sag in the Jizhong Depression. It is a large buried hill reservoir which composed of the carbonate strata of the Middle Proterozoic Wumishan Formation of the Jixian System, belonging to the fractured, bottom water, massive, low saturated reservoir[10]. Renqiu buried hill is divided into 10 oil groups, initially with a unified oil and water interface and a unified pressure system. This buried hill reservoir, being half-anticlines cut by several normal faulting, and the palaeotopograghy was steep in the west and north, low in the west and south. The formation, showing the characteristics of south old and north new, top surface morphology was in northeast direction, and the structure pattern was composed by four fault terraces and five hills. The buried fractures and faults are very developed. Due to the influence of multi-stage tectonic movement, the faults at different levels are overlapped with each other, and then the faults are further complicated. The faults are mainly developed in the northwest, north-east, north-north-east and nearly east-westward.

3.1. Fracture intensity model

According to the results of fractured logging, the reservoirs in the study area can be classified into four types: fracture development layers, with high porosity, indicated by the code "1"; fractures are relatively developed, with the code "2"; micro-fracture development, with the code "3"; other non-reservoir layer, that is, the fracture is not developed, with the code "0" to represent. In the process of fracture modeling, it is necessary to determine the probability distribution curve of each layer in the vertical direction first, as a constrained condition of the simulation process. By calculating and fitting the variogram of different levels of reservoirs, the author established the reservoir level model by sequential indication simulation method (Figure 1), the fractures are well developed. Through the outcrop and field geological observation, it is concluded that the frequency of tectonic fractures is related to the tectonic sites and faults, the closer to the faults, the more intensive the fractures are, and the fracture density decreases with the increase of the fault distance. In order to predict the density of fractures more rationally, this paper constructs the fault distance constraint model based on the fault distribution model, and predicts the fractures density as the auxiliary variable (Figure 2).

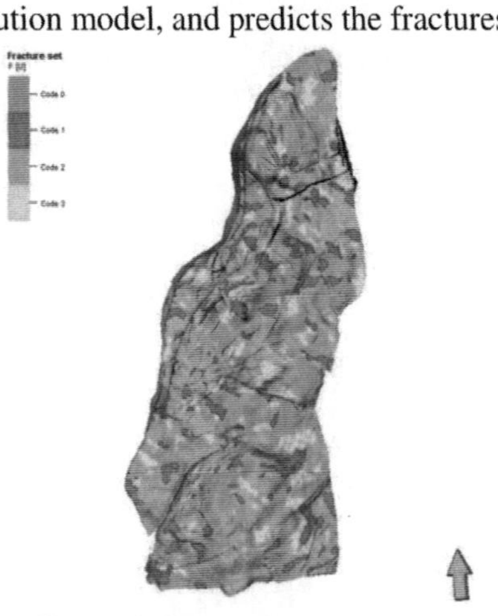

Figure 1. Reservoir level model.

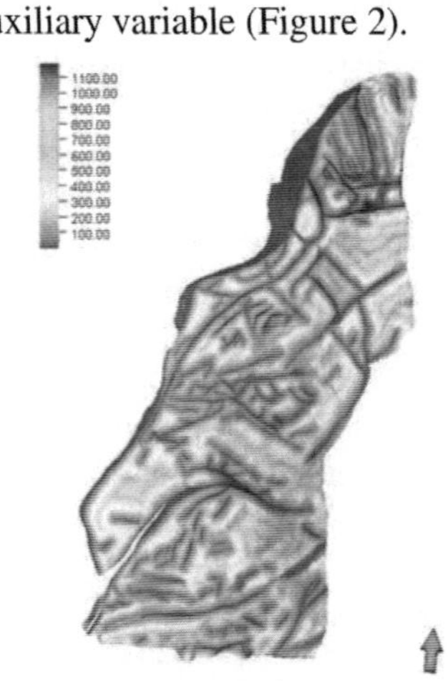

Figure 2. Fault distance analysis model.

Under the constraint of the reservoir level model, the fracture distance model is used as the second variable. Assigned the fracture density through the reservoir level, and then established the fracture intensity model by the sequential Gaussian simulation method (Figure 3).

3.2. Determination of simulation parameters for equivalent fracture modeling

In order to determine the relevant parameters of the equivalent fracture modeling and the effectiveness of the method, a small work area was selected to carry out the simulation test. The test area is 5Km^2, and the formation thickness is 400m. The structure is relatively simple. Taking the fracture density of 2 stripes per meter as an example, reduced it to 1 stripes per meter, 0.5 stripes per meter, 0.1 stripes per meter, 0.01 stripes per meter, establishing the discrete fracture network model, fractures porosity and permeability model, and then statistics calculating time. The statistical results are shown in Table 1.

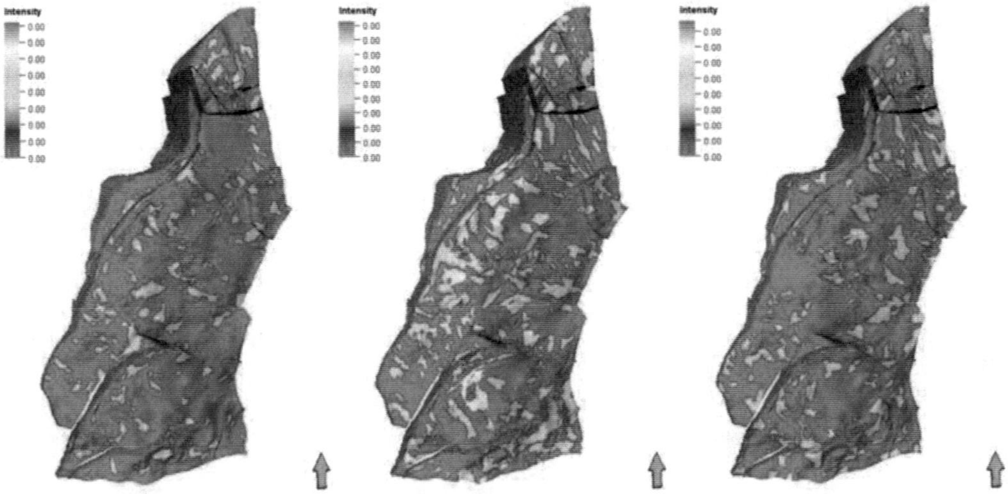

First grade Second grade Third grade

Figure 3. Fracture intensity model.

Table 1. Parameter settings of fracture modeling.

Mean fracture density (stripes per meter)	0.01	0.1	0.5	1	2
Fracture quantity	4423	150760	1075110	2511160	8861130
Mean fracture width (m)	0.003	0.0018	0.0008	0.0005	0.0004
Mean fracture length (m)	300	200	100	80	50
Mean fracture height (m)	75	50	20	10	8
Mean fracture porosity (%)	0.15	0.13	0.15	0.14	0.16
Mean fracture permeability (md)	43132	43547	42924	44511	42676
Time (min)	0.08	9	28	59	129
Computer configuration	Memory 16GB CPU:Intel(r) Core(TM)i7-4790 CPU@3.6GHz				

According to the results of the statistics in the table, it can be seen that by increasing the width, height and length of fracture, at the same time decreasing its density, can keep the similar porosity and permeability after upscaling discrete fracture network model. The area of the reservoir model of Wumishan Formation reservoir in Renqiu buried-hill is 57Km^2, the thickness of the formation is more than 2200 meters, and the average fractures density is about 2 Stripes per meter. Through simple conversion, we need to simulate more than 500 million fractures, and one simulation will takes for more than 120 hours. Due to memory constraints and other reasons in the actual simulations, the calculation time is not a simple linear increase, and then considering the stochastic simulation need to simulate multiple realizations, the traditional modeling method cannot meet the needs of practical work. By using the equivalent fractures modeling method, the calculation and upscaling time of the fracture are reduced by about two orders of magnitude, which satisfies the needs of the actual research work and ensures the consistency of the key geological parameters before and after upscaling.

Previous studies have shown that the density of the first grade fractures in the study area is less than 0.1 stripes per meter, the fractures density of the second grade fractures is about 0.5 stripes per meter, and the fractures density of the third grade fractures is between 0.5-10 stripes per meter. Petrel provides three kinds of fracture density, namely volume density, line density and areal density.

In the study area, the length of the first grade fractures is between 10-100m, the length of the second grade fractures is between1-10m, the length of the third grade fractures is between 0.5-1m, and a single grid volume of the model is 50 * 50 * 2m.

The fracture density of model is calculated according to the above formula:

$$\frac{\text{Length of single fracture} * \text{Empirical density of fracture} * \text{Length of single grid}}{\text{Volume of single grid}} \quad (3)$$

Table 2 is the fracture characterization parameter settings before and after the equivalent. The density of first, second and third grade fractures are reduced, and the length, height and width of fractures are increased.

Table 2. Fracture characterization parameter settings before and after the equivalent

	Density(stripes per meter)		Length(m)		Height(m)		Width(m)	
	BEF	AFT	BEF	AFT	BEF	AFT	BEF	AFT
Fracture grade 1	0.01	0.001	10	200	2	20	0.001	0.008
Fracture grade 2	0.02	0.002	4	80	1	10	0.0005	0.004
Fracture grade 3	0.005	0.0005	0.5	10	0.1	1	0.0001	0.001

3.3. Discrete fracture network model

This paper uses the popular ant tracking technique to simulate the large-scale and medium scale fractures that can be identified by seismic data[11]. Firstly, the original seismic attribute data is processed by structure smoothing, chaos, variance etc, to enhance the discontinuity of seismic data in the space[12], then based on the analysis of fracture characteristics in study area, tracking the seismic attribute through the reasonable parameter settings of ant tracking, and then construct the ant attribute model. Finally, the large scale discrete fracture distribution model is established by deterministic modeling method. For small scale fractures, take the fracture intensity model as constraints of the enrichment degree of fracture development, using discrete fracture network modeling module of Petrel software[13], simulate the small scale fractures by equivalent fractures modeling method proposed in this paper. Setting different fracture extension length, fracture height, fracture width and other fracture parameters for different grades of fracture, simulate fractures in different grades and different layers respectively. At the end of the simulation of fractures at various scales, the whole fractures model is constructed by a fracture network which composed of various fracture slices.

3.4. Fracture property model

Based on the fracture network model, the fracture network model is upscaled into fracture porosity and permeability model by using Petrel software. According to the actual data in the study area, using the Oda method, with a total area of fractures in single grid and different parameters of fractures as a basis to estimate the porosity and permeability, established the fracture porosity model and permeability model in three directions, which laid the foundation for the reservoir simulation.

4. Conclusions

a) In order to solve the problem that when the fracture intensity is great, generate a fracture network model for a large area will require long time, propose a method of equivalent fracture modeling. Equivalent multiple fractures to one fracture, the essence of the method is to reduce the overall density of fractures, increase the length, height, and width of the single fracture, it will greatly reduce the time required for fracture modeling, while ensuring pore and permeability simulation result is reasonable.

b) Setting different fracture extension length, fracture height, fracture width and other fracture parameters for different grades of fracture, simulate fractures in different grades and different layers respectively under the constraints of the fracture intensity property, described the distribution characteristics of fractures in study area finely.

c) Combined equivalent fracture modeling method with the ant tracking technology, established large-scale, medium-scale and small-scale discrete fracture network model for W1-W10 oil group in carbonate reservoir of Renqiu buried hills, The model described the geometry and the flow behavior of the fracture system in detail, providing basic support for late numerical simulation and remaining oil distribution study.

Acknowledgments

This project is supported by PetroChina Co Ltd Major Project, No.2014E-35-05 and the Innovation Group Project of Nature Science Foundation of Hubei Province, No.2016CFA024.

References

[1] Wang Jianhua. DFN model: A new modelling technology for fracture [J]. *Fault Block Oil & Gas Field*, 2008, **15**(**6**):55-58.

[2] Nie Yongsheng, Tian Jingchun and Wei Shengxiang, et al. Difficulties and treatments in three-dimensional geological modeling of fractural reservoirs [J]. *Petroleum Geology and Recovery Efficiency*, 2013, **20**(**2**):39-41.

[3] Xu Zhihua. Study on geological modeling Techniques for Discrete Fracture Network [D].*China University Of Petroleum* 2009

[4] Tong Henmao. Establishment of the fracture network model of analogous outcrop and its application in the buried hill reservoir in Renqiu [J]. *Journal of Xi'an Shiyou University (Naturnal Science Edition)*, 2006, **21**(**3**):17-22.

[5] Zhang Shujuan, Wang Yanbin and Liang Xingru, et al. Application of ant tracking technology in fracture prediction of carbonate buried hill reservoir [J]. *Fault-Block Oil & Gas Field*, 2011, **18**(**1**):51-54.

[6] Zhang Shujuan, Liu Dating and Yang Yuxiang, et al. The remaining oil distribution pattern and tapping direction of buried hill reservoir in Renqiu [J]. *Petroleum Geology and Engineering*, 2007, **21**(**5**):50-54

[7] Zheng Songqin, Yao Zhiliang. Discrete Fracture Network Stochastic Modeling [J]. *Journal of Oil and Gas Technology*, 2009,**31**(**4**):106-110.

[8] Ju Yan, Sun Xiongwei and Zhang Chengsen, et al. Multi-Approach Characterization and Distribution of Ordovician Carbonate Fractures in Tazhong Area, Tarim Basin [J]. *Xinjiang Petroleum Geology*, 2011, **32**(**3**):262-266.

[9] Mu Longxin, Zhao Guoliang and Tian Zhongyuan, et al. Study on reservoir fracture prediction [M]. *Petroleum Industry Press*, 2009.

[10] Yu Jiaren, Fan Zheren. A study on the carbonate reservoir rock of Renqiu buried-hill oilfield [J]. *Acta Petrolei Sinica*, 1981, **2**(**1**):57-68.

[11] Sun Le, Wang Zhizhang and Li Hanlin, et al. Application of fault tracking technology based on ant colony algorithm in Wuxia Area [J]. *Fault-Block Oil & Gas Field*, 2014, **21**(**6**):716-721.

[12] Hu Shaohua, Guo Bo and Lin Dongcheng,et al. 3D seismic data fine reservoir prediction in structural Fang-3 interpretation and Well Block [J]. *Oil Geophysical Prospecting*, 2010,**45**(**6**):909-913.

[13] Feng Jianwei, Chang Lunjie and Sun Zhixue, etc. Geological model and characteristics of discrete fracture network in tight sandstone gas reservoir constrained by multi-factors [J]. *Journal of China University of Petroleum (Edition of Natural Science)*, 2016, **40**(**1**):18-26.

Evaluation of Losses Of Cold Energy of Cryogen Products in The Transport Systems

Dmitry Uglanov[1, *], Dmitry Sarmin[1], Alexandra Tsapkova[1] and Yana Burdina[1]

[1]Samara University, Department Of Heat Engineering And Heat Engines, 443086, Moskovskoye shosse, 34, Samara, Russia

*Corresponding author: dmitry.uglanov@mail.ru

Abstract. At present, there are problems of energy saving in various areas of human life and in power complexes of industrial plants. One possible solution to the problem of increasing energy efficiency is the use of liquefied natural gas and its cold energy. Pipelines for fuel or gas supply in cryogen supply systems have different length depending on the mutual position of storage and cryogen consumption devices relatively to a start construction. Cryogen supply and transport systems include a lot of fittings of different assortment. Reservoirs can be installed on different elevation points. To reduce heat inleak and decrease cold energy of cryogen product different kinds of thermal insulation are used. Cryogen pipelines provide required operation conditions of storage and gasifying systems. The aim of the thermal calculation of cryogen transport and supply systems is to define the value of cryogen heat. In this paper it is shown values of cryogen temperature rise due to heat inleaks at cryogen's transfer along transport systems for ethane, methane, oxygen and nitrogen were calculated. Heat inleaks also due to hydraulic losses were calculated. Specific losses of cold energy of cryogen product for laminar and turbulent flow were calculated. Correspondences of temperature rise, critical pipeline's length and Reynolds number were defined for nitrogen, argon, methane and oxygen.

1. Introduction

At present, there are problems of energy saving in various areas of human life and in power complexes of industrial plants[1,2]. One possible solution to the problem of increasing energy efficiency is the use of liquefied natural gas and its cold energy [3].

Some aspects of cryogen supply and transportation are investigated in works of A.M. Arharov, I.D. Kunis [4], N.V. Filin [5], M.P. Malkov [6]. On the ground of their proceedings short description of their works can be made.

Pipelines for fuel or gas supply in cryogen supply systems have different length depending on the mutual position of storage and cryogen consumption devices relatively to a start construction. Cryogen supply and transport systems include a lot of fittings of different assortment. Reservoirs can be installed on different elevation points. To reduce heat inleak and decrease cold energy [7, 8, 9, 10] of cryogen product different kinds of thermal insulation are used.

Thermal and hydraulic calculations should be made to design cryogen supply and transport systems.

Cryogen pipelines provide required operation conditions of storage and gasifying systems. Cryogen pipe is a coaxial two-walled pipe with a screen-vacuous insulation. Cryogen pipeline sections are welded.

A right-of-way of the cryogen pipeline is assembled according to the location with the following elements:

- sections of cryogen pipelines;
- compensation elements (compensators, metal hoses); reducing fitting (joint sleeves, lens etc.);
- adsorptive sections;
- vacuum ports for connection to evacuation and vacuum control equipment;
- supports, membrane protectors, safety grounding devices.

The aim of the thermal calculation of cryogen transport and supply systems is the definition of the value of cryogen heat. There are two reasons for the calculation:

- maintenance of the cryogen temperature that a consumer needs;
- single-phase condition of the cryogen.

When calculating, the system is considered to be absolutely preliminary chilled to the temperature of a liquid product.

2. Calculation methods

The increase of the cryogen temperature is given by the expression:

$$\Delta T = \frac{\Sigma Q}{c_p G}. \quad (1)$$

Heat of temperature rise is comprised of:

- temperature rise due to the heat inleak from the environment;
- temperature rise due to the hydraulic friction in pipelines and the throttling effect.

Thus, temperature rise of the cryogen can be defined by the following expression:

$$\Delta T_S = \Delta T_{o.c.} + \Delta T_{г}, \quad (2)$$

where ΣQ – overall heat inleak to the system from the environment, V; c_p – product thermal capacity, J/kg·K; G – consumption of fuel component, kg/s.

Specific overall heat inleak to the cryogen product for 1 m is defined by the following expression:

$$q_\Sigma = q_l + q_{mp}, \quad (3)$$

where q_l – heat inleaks through thermal insulation layer and pipelines (sleeves), V/m; q_p – hydraulic losses in pipelines, V/m.

3. Heat inleaks calculation

For computational convenience a filling system is divided into separate sections. Heat inleaks are defined for every section and system's element separately, and then they are summed up. Formulae for calculation of heat inleaks through some filling system's elements are given below.

Heat inleaks through thermal insulation layer and sleeves are defined as

$$q_l = K_l \cdot (T_{o.c.} - T_\kappa) = \frac{T_{o.c.} - T_\kappa}{\frac{1}{\alpha_1} + \frac{d_1}{2\lambda} ln\frac{d_2}{d_1} + \frac{d_1}{2\lambda_{из}} ln\frac{d_{из}}{d_2} + \frac{d_1}{\alpha_2 d_{из}}}, \quad (4)$$

where T_{env} – environment temperature, K; T_κ – component temperature, K; α_1 – heat-transfer from the wall to the cryogen product coefficient, W/(m^2·K);α_2 – heat-transfer from environment to the wall coefficient, W/(m^2·K), that equals 10 W/(m^2·K) according to [1]; d_1 *and* d_2 – inside and outside diameter of the pipe, m; d_{ins} – outer insulation diameter, m; λ_{ins} – thermal conductivity coefficient of the insulation layer, W/(m^2·K):

- for a powder-vacuous insulation $\lambda_T = 2{,}1 \cdot 10^{-3}$ V/ (m·K);
- for a screen-vacuous insulation $\lambda_T = (1{,}1...1{,}2)\cdot 10^{-3}$, V/ (m·K);

- for a non-vacuum insulation of ATIMS -10 λт = 4,2·10-3 V/(m·K), and ATIMS-3-15 λт = 3,5·10-3 V/(m·K) (more accurate data of coefficients depending on pressure and number of insulation layers are given in references);

Heat-transfer coefficient from the wall to the cryogen is defined by the value of Nu criterion:

$$\alpha_1 = \frac{Nu \cdot \lambda_{жс}}{d_1} = \frac{0{,}21 \cdot Re_{жс}^{0{,}8} \cdot Pr_{жс}^{0{,}43} \cdot \lambda_{жс}}{d_1}. \quad (5)$$

Nusselt criterion is defined by the flow regime. At laminar fluid flow, $Re_{f,d} < 2{\cdot}10^3$, to define average heat transfer coefficient the following formula is suggested:

$$\overline{Nu_f} = 0{,}15 \cdot Re_f^{0{,}8} \cdot Pr_f^{0{,}43} \cdot Gr_f^{0{,}1} \cdot \left(\frac{Pr_f}{Pr_W}\right)^{0{,}25}. \quad (6)$$

At developed turbulent flow,$Re_f > 10^4$, the calculation formula is as follows:

$$\overline{Nu_f} = 0{,}21 \cdot Re_f^{0{,}8} \cdot Pr_f^{0{,}43} \cdot \left(\frac{Pr_f}{Pr_W}\right)^{0{,}25}. \quad (7)$$

The definiens are a pipe's diameter *d*, defining temperature t_f, average fluid velocity *W*.

In cryogen supply systems heat inleaks can appear in the following elements:

- Heat inleak through bayonet joints of cryogen pipelines;
- Heat inleak along a butt joint through a bellows;
- Heat inleak along supporting structures of pipe utilities (pin supports);
- Heat inleak along fittings.

After the calculation of heat inleaks of separate elements and system sections, the results are presented in tables. To consider all losses at totaling the values, coefficient that equals 1.3-1.5 is used.

As the calculations and experience of working with cryogen pipelines show, the basic heat of cryogen fluid, transported along supply system, is conditioned by temperature heat due to heat inleaks from environment. Heat of hydraulic losses is somewhat smaller.

The temperature of the liquid component at the inlet to the cryogenic systems:

$$\Delta T_{f.j.} = \Delta T_{res} + \Delta T_{heat}, \quad (8)$$

where ΔT_{res} – change of temperature in the tank of the filling system, K;

ΔT_{heat} – heat value of hydraulic losses, K.

According to references about pipelines, designed and produced in company "Cryogenmash"[11], calculation of specific inleaks per meter run of pipelines of different diameter is made. The values of specific heat inleaks are presented in the table 1 and in the figure 1.

q, W/m

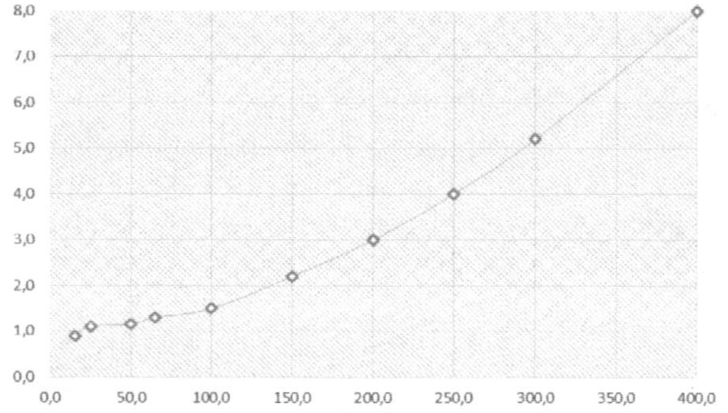

Pipe diameter, mm

Figure 1. Correspondence of a specific heat inleaks per 1 meter of a pipeline and its inside diameter

4. Heat inleaks calculation caused by hydraulic losses

Figure 2 shows transport scheme of cryogen liquid at a distance. In that system heat, supplied to the cryogen in a pump, is given to fluid under low pressure in a heat exchanger, and discharged fluid under a pressure (higher then bubble pressure) is supplied to the pipeline. The problem was considered with fluid approaching a valve at saturation.

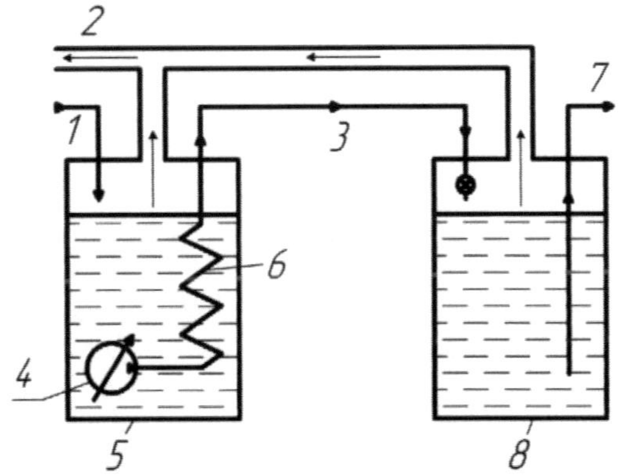

Figure 2. Transport system of cryogenic liquid: 1— fluid flow; 2 — steam return or release; 3 — pipeline; 4 — pump; 5 — pump reservoir; 6 —heat exchanger; 7 —to a consumer; 8 — distant storage.

Fluid heat due to hydraulic losses can be found by the following formula:

$$\Delta T_{h.} \cong \frac{\Delta P}{\rho \cdot c_p}, \tag{9}$$

where ΔP – pressure losses, MPa (defined by hydraulic calculation of the system); ρ – air density, kg/m^3.

Cryogen pressure losses at cryogen's transposition inside a pipeline are defined by flow regime, specifically by Reynolds number (*Re*). While flow velocity is defined as follows:

$$w = \frac{Q}{S} = \frac{4 \cdot Q}{\pi \cdot d^2}, \tag{10}$$

where Q – cryogen volume flow rate; S – line flow area.

Pressure losses on friction in general are defined as follows:

$$\Delta p = \lambda \cdot \frac{l}{d} \cdot \frac{\rho w^2}{2}, \tag{11}$$

where λ – specific friction coefficient; ρ – cryogen density; l – pipeline section's length.

Correlation for pressure losses on friction per one meter run calculation is defined as follows:

$$\Delta p_l = \frac{\lambda}{d} \cdot \frac{\rho w^2}{2}, \left[\frac{Pa}{m}\right]. \tag{12}$$

Specific losses of low temperature heat of cryogen flow per 1 meter of a pipeline are defined as follows:

$$q_l^k = \frac{\lambda}{d} \cdot \frac{w^2}{2}, \left[\frac{J}{kg \cdot m}\right]; \tag{13}$$

$$q_l^k = \frac{\lambda}{d} \cdot \frac{16Q^2}{2d^4} = \frac{8\lambda Q^2}{\pi^2 d^5}; \tag{14}$$

$$q_l^k = \frac{8\lambda G_m^2}{\pi^2 d^5 \rho^2}. \tag{15}$$

The last correlation has mass flow:

$$G_m = \rho \cdot Q. \tag{16}$$

Friction coefficient λ is defined by a flow regime. For a laminar cryogen flow:

$$\lambda = \frac{64}{Re} = \frac{64 \cdot \upsilon}{w \cdot d} = \frac{64 \cdot \upsilon \cdot \pi d^2}{4 \cdot Q} = \frac{64 \cdot d^2\, \upsilon \cdot \rho \pi}{4 \cdot G_m}. \tag{17}$$

For turbulent cryogen flow:

$$\lambda = \frac{0{,}3164}{Re^{0.25}} = \frac{0{,}3164 \cdot \upsilon^{0.25}}{(w \cdot d)^{0.25}} = \frac{0{,}3164 \cdot \upsilon^{0.25}}{\left(\frac{4Q \cdot d}{\pi \cdot d^2}\right)^{0.25}} =$$

$$= \frac{0.3164 \cdot \pi^{0.25} \cdot d^{0.25} \cdot \upsilon^{0.25}}{4^{0.25} \cdot Q^{0.25}}. \quad (18)$$

Considering calculation expressions for definition of friction coefficient, specific energy losses of cryogen flow are defined as follows:

For a laminar flow:

$$q_{l_{lam}}^{k} = \frac{8 \cdot 64 d^2\, \upsilon \cdot \rho \cdot \pi \cdot G_m^2}{4 \cdot G_m \cdot \pi^2 \cdot d^5 \cdot \rho^2} = \frac{128 \cdot \upsilon \cdot G_m}{\pi \cdot d^3 \cdot \rho}, \quad (19)$$

For a turbulent flow:

$$q_{l_{turb}}^{k} = \frac{8 \cdot 0{,}3164 \cdot \pi^{0{,}25} d^{0{,}25} \upsilon^{0{,}25} \cdot G_m^2}{\pi^2 d^5 \cdot 4^{0{,}25} \cdot G_m{}^{0{,}25} \rho^2} = 0{,}24 \frac{\upsilon^{0{,}25} \cdot G_m^{1{,}75}}{d^{4{,}25}\, \rho^{1{,}75}}. \quad (20)$$

In general losses of cold energy due to friction will be defined by the following expression:

$$g_l = f(G_m, \upsilon, d, \rho\,). \quad (21)$$

Table 1. Heat of cryogen evaporation

Fluid	Empirical expression $r = r(p)$, [kJ/kg]
Nitrogen	$r = 200{,}1 \cdot e^{-0{,}02p}$
Argon	$r = 160{,}2 \cdot e^{-0{,}01p}$
Oxygen	$r = 212{,}9 \cdot e^{-0{,}01p}$
Methane	$r = 516{,}2 \cdot e^{-0{,}02p}$

Critical pipeline's length, when cryogen gasifies, is defined by the following expression, considering heat of cryogen evaporation, depending on pressure $r = r(p)$ (table 1):

$$l_{cr} = \frac{r \cdot G_m}{q_\Sigma}. \quad (22)$$

Based on given expressions heat flows and critical cryogen pipelines' length were calculated. Using this methodology it has been calculated the dependence of Specific overall heat inleak, the dependence of the temperature change of the liquid component and the dependence of the critical cryogen pipelines length on the Reynolds number. The results are shown at figures 3-5.

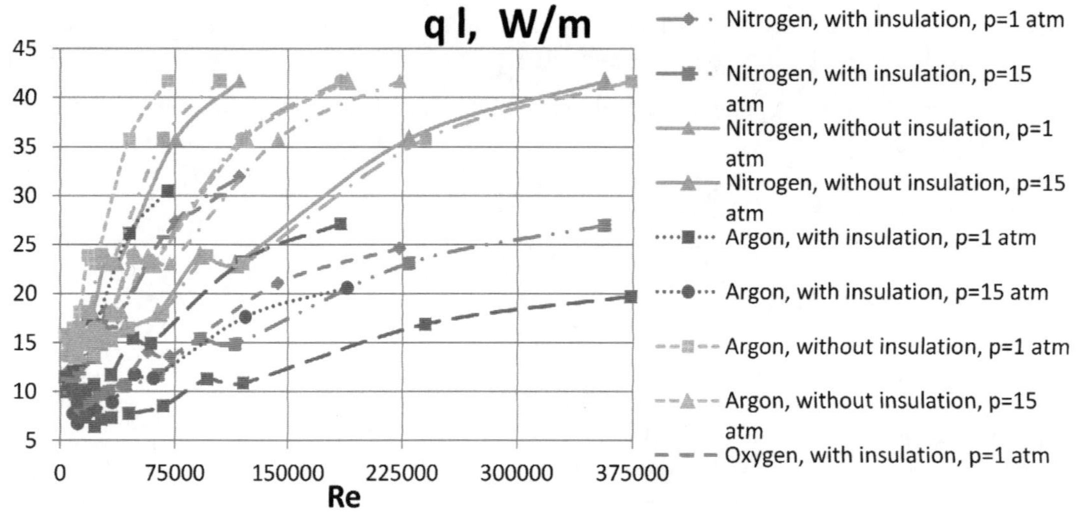

Figure 3. The dependence of Specific overall heat inleak to the cryogen product for 1 m t on the Reynolds number

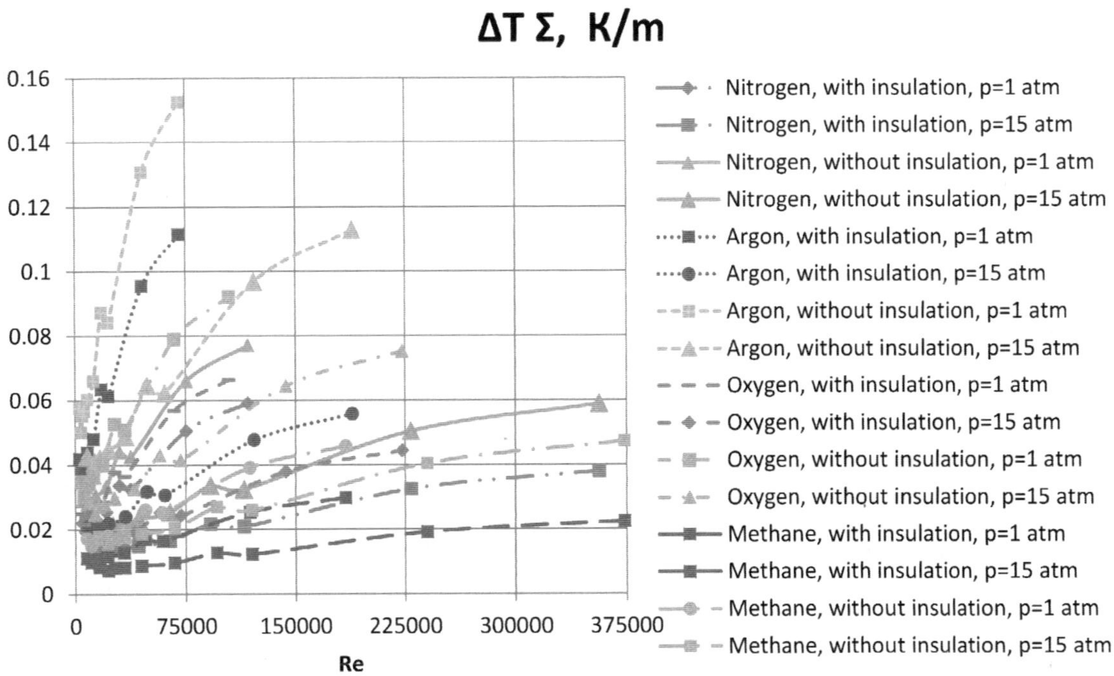

Figure 4. The dependence of the temperature change of the liquid component on the Reynolds number

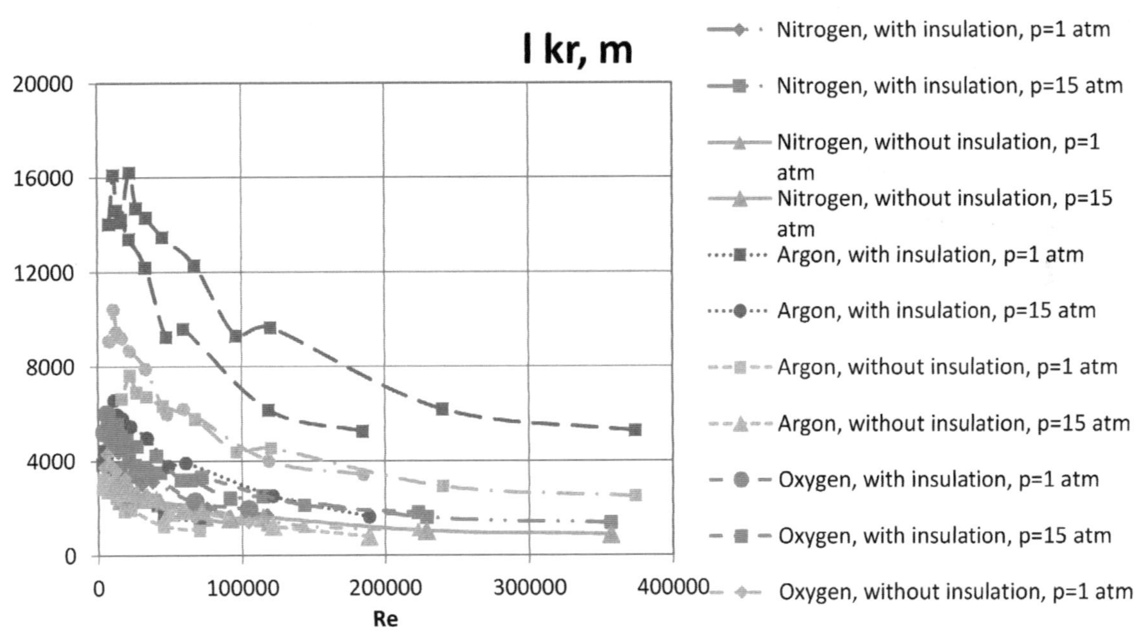

Figure 5. The dependence of the critical cryogen pipelines length on the Reynolds number

5. Conclusions

In the result of the work:

1) Values of cryogen temperature rise due to heat inleaks at cryogen's transfer along transport systems for ethane, methane, oxygen and nitrogen were calculated.

2) Heat inleaks due to hydraulic losses were calculated.
3) Specific losses of cold energy of cryogen product for laminar and turbulent flow were calculated.
4) Correspondences of temperature rise, critical pipeline's length and Reynolds number were defined for nitrogen, argon, methane and oxygen.

References

[1] Kerdphol, T., Qudaih, Y., Mitani, Y.Optimum battery energy storage system using PSO considering dynamic demand response for microgrids.(2016) International Journal of Electrical Power and Energy Systems, 83, pp. 58-66.

[2] Kakumoto, Y., Koyamatsu, Y., Shiota, A., Qudaih, Y., Mitani, Y. Application of Geographic Information System to Power Distribution System Analysis (2016) Energy Procedia, 100, pp. 360-365.

[3] Burtsev, S.A., Karpenko, A.P., Leontiev, A.I. A method for distributed production of liquefied natural gas at gas-distribution stations. (2016) High Temperature, 54 (4), pp. 573-576.

[4] A.M. Arkharov, I.D. Kunis. *Cryogenic refueling systems of launch rocket and space complexes*, publishing house MSTU. NE Bauman, 2006.

[5] N.V. Filin, A.B. Bulanov. *Liquid cryogenic systems*, Leningrad, "Mechanical Engineering", 1985.

[6] M.P. Malkov, *Handbook on the Physical and Technical Basics of Cryogenics*, Moscow, Energoatomizdat, 1985.

[7] G. Zhang, W. Xu, Y. Yang, D. Zhang, Energy Procedia, *Utilization of LNG cryogenic energy in a proposed method for inlet air cooling to improve the performance of a combined cycle*, **Volume 61**, p. 899-903, (2014).

[8] Y. Xiong, P. Luo, B. Hua, Energy Procedia, *A novel CO2-capturing natural gas combined cycle with LNG cold energy utilization.***Volume 61**, pages 899-903, (2014).

[9] E.V. Blagin, A.I. Dovgyallo, S.O. Nekrasova, D.V. Sarmin, D.A. Uglanov, Applied Thermal Engineering *Estimation of the energy efficiency of cryogenic filled tank using in different system and devices*. 2016. **Volume 101**, 25 May 2016, Pages 537-544

[10] H. Sun, H. Zhu, F. Liu, H. Ding, Energy, *Simulation and optimization of a novel Rankine power cycle for recovering cold energy from liquefied natural gas using a mixed working fluid*, **Volume 70**, pages 317-324, (2014).

[11] http://www.cryogenmash.ru/tpl/Catalogue_CM_2014.pdf

The Vulnerability Formation Mechanism and Control Strategy of the Oil and Gas Pipeline City

Y L Chen[1, 2] and L Han[2, 3,*]

[1]Institute of Environmental and Chemical Engineering, Jiangsu University of Science and Technology, Zhenjiang, Jiangsu, China
[2]Department of Urban Water Environmental Research, Chinese Research Academy of Environmental Sciences, Beijing, China
[3]State Key Laboratory of Environmental Criteria and Risk Assessment, Chinese Research Academy of Environmental Sciences, Beijing 100012, China

*hanlu@craes.org.cn; 1194772844@qq.com

Abstract. Most of the pipelines of oil and gas pipelines in our country have been for more than 25 years. These pipes are buried underground and was difficult to daily test. In addition, it was vulnerable to environmental, corrosion and natural disasters ,So there is a hidden nature of accidents. The rapid development of urbanization, population accumulation, dense building and insufficient safety range are all the reasons for the frequent accidents of oil and gas pipelines. Therefore, to appraise and know the safe condition of the city various regions oil and gas pipelines is vital significant.In order to ensure the safety of oil and gas pipeline city, this paper defines the connotation of oil and gas pipeline city vulnerability according to the previous research on vulnerability. Then from three perspectives of environment ,structure and behavior, based on the analytical paradigm of "structure—vulnerability conduct—performance" about oil and gas, the influential indicators of vulnerable oil and gas pipelines were analysed ,the vulnerability mechanism framework of Oil and gas pipeline city was also constructed. Finally ,the paper proposed the regulating strategy of the vulnerability of the oil and gas pipeline city to decrease its vulnerability index, which can be realize the city's vulnerability evaluation and provides new ideas for the sustainable development of the city.

1. Introduction

Oil and gas pipelines in China were expected to increase to meet growing demand. Oil and gas pipelines in our country have undergone four stages initial development (1958-1969), rapid development (1970-1987), steady development (1988-1995) , accelerated development(1996 - present)[1].Most oil and gas pipelines in China were more than 25 years, there exist many safety problems such as leakage of products, sick operation and serious corrosion etc. Separately, it can't be repaired or replaced in time. The average rate of oil and gas pipeline accidents in China was 3 times per 1000 km/year, well above the US 0.5 times per 1000 km/year and Europe's(0.25 times per 1000 km/year), public information shows. Incomplete statistics show that there have been more than 1000 oil and gas accidents in China since 1995[2]. As soon as the oil and gas pipeline leaked or breaked, it will cause significant fire accident and even malignant consequences such as explosion, poisoning, environmental pollution, especially in densely populated areas, serious casualties and great economic losses tend to appear. In 2002,Six people were killed and five injured in a blast which was attributed to the bulid up of gas in Daqing. In 2006,A "1.20" explosion in the gas station of the Fujia gas station in the southwestern oil and gas field was led to a bad result:10 deaths,3 serious injured and 47 slight injured. "The great accident of the 11.22 Donghuang oil pipelines" in 2013, which killed 62 people and 136 was injured.

As a new term for the study of security, vulnerability is an important bridge between disaster and risk research and also is an important comprehensive indicator of system security[3].Based on many dangers to the surrounding cities of oil and gas pipelines, it is of practical significance to study the formation mechanism of oil and gas pipelines city vulnerability to the possibility of scientific prevention and reduce the occurrence of disaster.

2. Oil and gas pipelines vulnerability

The term of vulnerability was applied to the field of geosciences and natural disaster research at the beginning, which was interpreted as the extent of loss caused by some elements of risk disasters[4]. In the 1990s, vulnerability was associated with resilience, marginality, sensitivity, and adaptability[5-7].The concept of vulnerability has different expressions in different research fields, but there are certain commonalities: the instability of the system itself. Internal characteristic of the system was the most important and direct factor affecting the fragility. External interference can affect the size of vulnerability by the internal characteristics of system, and then be reflected in the sensitivity and response of the system to the interference. Vulnerability not only reflects the current state of the system, but also includes the risk problem, which presents the spatial and temporal pattern of vulnerability.

As to the analysis above, vulnerability was reviewed as a status property in this paper, that is, the oil and gas pipeline city system exhibits a high degree of sensitivity under the external disturbance and lacks of response capability, cause the system to the extent of damage. The vulnerability of the system is usually inversely correlated to security. The vulnerability of the system went down when security on it was improved. Subject to its sensitivity and instability, under the effect of disturbance of different factors, the oil and gas pipelines city system shows a high degree of complexity.

3. Formation mechanism of oil and gas pipelines city

Factors affecting the vulnerability of oil and gas pipelines city can be divided into three levels: environment、structure and behavior, and its vulnerability is changed by the interaction between direct, indirect or influencing factors. The vulnerability of oil and gas pipelines in this paper was a negative performance. City, the industrial chain or the formation of the capital structure tend to have a long time, formed its unique structure of rigid. This feature in the face of the urban resource constraints, the adverse effects such as changes in oil and gas pipelines, will show its effect, which reflects the side effects of oil and gas pipelines city. Taking oil and gas pipelines as a system, a combination of society, economy and resources environment have reflected its exposure, sensitivity and adaptability, etc. In addition, the sensitivity is the property of the oil and gas pipelines city itself, which is determined by its internal structure. The size reflects the response of the oil and gas pipeline city system to the internal and external changes, which may be positive or negative. The response capacity was reflected in oil and gas pipelines city system, that displayed against a variety of characteristics of the adverse impact when the internal or external disturbance occurred. The changes that system faced were concealment or controllable, and cope with these changes rely mainly on the function of the system itself. Besides, not only the adaptation of the oil and gas pipeline city system to external changes included, but also the variability of the external changes, namely active coping and reactive coping[8,9]. The adaptive value is greater, the stronger the adaptability of the system, the less vulnerable the system is. Resilience refers to the bear, adaptation, or the ability to recover the system after the internal or external disturbance happens, that is, the time and energy required for the system to recover from the disaster. The model of formation mechanism of oil and gas pipelines city vulnerability was constructed, as shown in Fig.1.

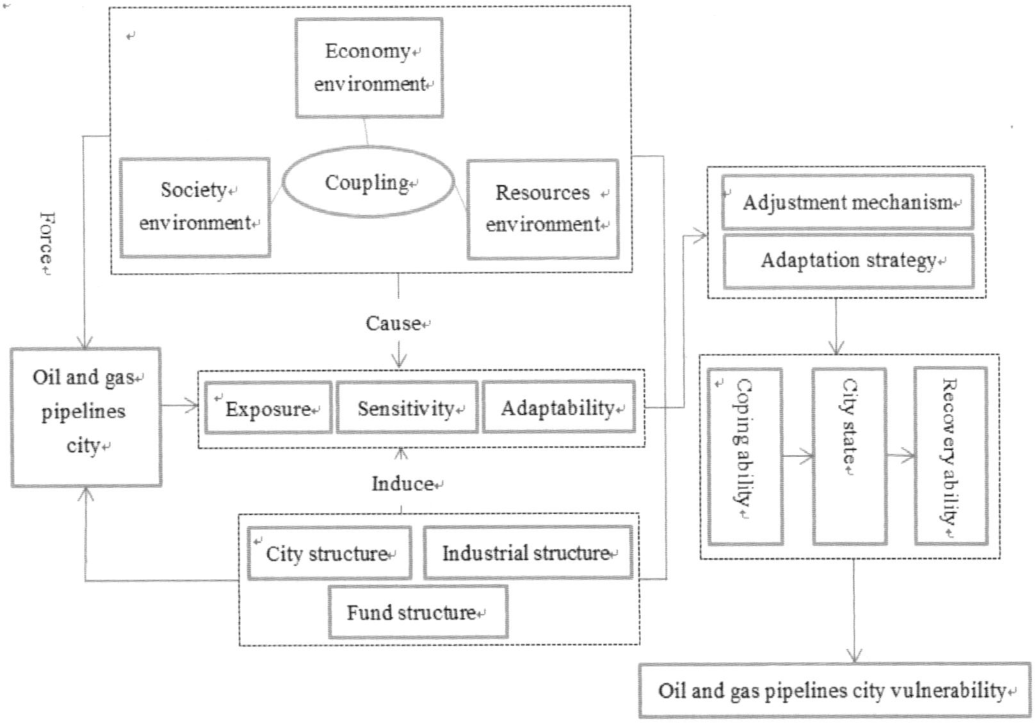

Figure 1. The model of formation mechanism of oil and gas pipelines city vulnerability

3.1 City vulnerability

a) Economic factors: Mainly reflects the economic strength of the oil and gas pipeline city, and directly related with the coping ability. The higher the level of economic and social development, the stronger the economic strength, oil and gas pipeline city on the internal or external interference resistance is stronger.

b) Society factors: The impact of social factors on vulnerability includes population density, age structure, population quality and other population factors, medical and health conditions, culture and education and other public services were also considered. In general, for areas where the density of the city is large, the more likely it is to lead to unexpected losses, the higher the sensitivity, the lower the safety.

c) Resources environment factors: In the process of urbanization, Economic growth is bound to sacrifice environmental quality for a certain price. Environmental pollution problems in resource-constrained settings have increasing prominent.

d) Oil and gas pipelines factors: Oil and gas pipelines vulnerability often manifest as the complexity of its surroundings. Moreover, it was susceptible to population density, soil geological conditions and so on. The characters of centralized distribution, large amount and medium kinds will increase the vulnerability of the oil and gas pipeline system.

3.2. The vulnerability of emergency capability

The emergency capability has a critical role to play in the occurrence of disasters, that is the "anti-sensitivity" of the system. On the whole, coping capacity presents two states in the event of a disaster: one is the vulnerability of oil and gas pipelines. By taking measures to decrease the potential of oil and gas pipelines, reduce the exposure of oil and gas pipelines or minimize damage. Another is city vulnerability .Using the coping capacity to weaken the vulnerability factor, that is, to enhance the ability of the city itself to counter the adverse effects, and then to reduce the vulnerability of the system being evaluated.

4. The control measures and suggestions of vulnerability on oil and gas pipelines city

1) Increase the awareness of security knowledge
2) Leaking, once happening to pipeline, will easily lead to fire or explosion and cause damage to the surrounding environment, especially in complex areas such as population and buildings concentrated regions. Therefore,it is important for residents to realize the harmfulness of pipelines and raise their safety concerns. Try to aviod some adversely behaviors to the pipes, including the third-party damage, unlawful tying, etc. Besides that, we can use the strength of the residents to supervise construction enterprises so as to reduce error operation,which can help decrease the vulnerability of system effectively.
3) Establish linkage mechanism for government, enterprises and community
4) "11 • 22 Donghuang oil pipelines leakage explosion" was identified as accident caused by negligence. Based on this, it is necessary to establish a co-management mechanism that guided by the government, and enterprises, community participated in. On the one hand, to strengthen the national safety production-related laws or regulations while the government as the main force. On the other hand, enterprises and community residents are brought into the management area to form a system of mutual supervision.
5) Adjust industrial structure and implement sustainable management
6) Taking the oil and gas pipelines city as a system, raise the proportion of the tertiary industry to reduce the negative effects on the environment caused by lower industrial structure. Correctly handle the relationship between the inside and outside of the system, and to predict the potential changes in the system and possible consequences happend. To coordinate the economic and environmental development through the implementation of sustainable management, so as to improve the safety of the oil and gas pipelines city.

5. Conclusion

1) Based on the concept of vulnerability related concepts in different fields, the concept of vulnerability of oil and gas pipelines is proposed. In this paper, vulnerability was reviewed as a status property, that is, the oil and gas pipeline city system exhibits a high degree of sensitivity under the external disturbance and lacks of response capability, cause the system to the extent of damage.
2) Vulnerability is the inherent attribute of oil and gas pipeline city. From the perspective of the concept of oil and gas pipelines vulnerability, this paper divided it into two parts: city vulnerability and emergency capability vulnerability. Then the model of formation mechanism of oil and gas pipelines city vulnerability was contructed and analyzed.
3) The paper put forward the control measures and suggestions of vulnerability on oil and gas pipelines city, which can help to identify the root of vulnerability and find the weak links of the evaluation system. It will provides scientific theoretical basis for the oil and gas pipeline safety management to ensure the safety of the oil and gas pipelines.

References

[1] Yao W,Jiang C L and Cong J H 2010 *J.*Comprehensive safety risk precontrol of the pipelines. *China Petroleum Enterprise.***(12)**.89-93
[2] Qi A H 2009 *J.* The present situation and problem analysis of oil and gas pipeline transportation in China. *International Petroleum Economics*.**17(12)**,57-59
[3] Zhao X W, Chai J S 2011 *J.* Study on indicators of vulnerability assessment for city buried gas pipeline system. *Journal of Safety Science & Technology*.**7(7)**,92-96.
[4] Timmerman P 1981 *J.* Vulnerability, Resilience and the Collapse of Society. A Review of Models and Possible Climatic Applications.*Institute for Environmental Studies*, University of Toronto
[5] Huang L L, Yao A L and Xian T 2014 *J.* Research on risk assessment method of oil&gas pipelinewith consideration of vulnerability.*China Safety Science Journal*. **24(7)**.93-99

[6] Krellenberg K,Link F and Welz J 2014 *J.* Supporting local adaptation: The contribution of socio-environmental fragmentation to urban vulnerability.*Applied Geography*. **55(12)**.61-70

[7] Gallopín G C 2003.Handbook for Estimating the SocioEconomic and Environmental Effects of Disasters.*Economic Commission for Latin America and the Caribbean*. Mexico.45-56

[8] Yohe G, Tol R.S J 2001 *J.* Indicators for social and economic coping capacity- moving toward a working definition of adaptive capacity.*Global Environmental Change*. **12(1)**.25-40

[9] Feng L I 2013 *J.*The Frangibility of Tourism Economy:Definition,Formation Mechanism and Frame Analysis. *East China Economic Managemen*.**27(03)**.76-81

A Low Cost Approach to the Design of Autopilot for Hypersonic Glider

Wang Liang[1], Zhang Weihua[1], Peng Ke[1] and Wang Donghui[1]

[1]School of Aerospace Science and Engineering, National University of Defense Technology, Deya Road 109, Kaifu District, Changsha, China

*Corresponding author: liangwang27@hotmail.com

Abstract. This paper proposes a novel integrated guidance and control (IGC) approach to improve the autopilot design with low cost for hypersonic glider in dive and pull-up phase. The main objective is robust and adaptive tracking of flight path angle (FPA) under severe flight scenarios. Firstly, the nonlinear IGC model is developed with a second order actuator dynamics. Then the adaptive command filtered back-stepping control is implemented to deal with the large aerodynamics coefficient uncertainties, control surface uncertainties and unmatched time-varying disturbances. For the autopilot, a back-stepping sliding mode control is designed to track the control surface deflection, and a nonlinear differentiator is used to avoid direct differentiating the control input. Through a series of 6-DOF numerical simulations, it's shown that the proposed scheme successfully cancels out the large uncertainties and disturbances in tracking different kinds of FPA trajectory. The contribution of this paper lies in the application and determination of nonlinear integrated design of guidance and control system for hypersonic glider.

1. Introduction

Hypersonic flight has the characteristic of great fast velocity and long distance, thus it has attracted great interests of researchers in recent years. However, hypersonic flight dynamics is associated with coupled nonlinearities and aerodynamics uncertainties, and the guidance and control system requires both good transient performance and high steady accuracy, so its design is a challenging task.

Compared to the conventional design of guidance and control system, the integrated method benefits a lot for its low cost and high accuracy [1, 2]. The rationale of integrated method is to develop the guidance and control system based on the overall nonlinear system dynamics. This paper has further consideration of actuator dynamics based on the integrated guidance and control scheme, because large chattering and peak phenomenon always appears in the control input when the system dynamics contains large nonlinearity and uncertainty. Basically, state feedback back-stepping technique is a powerful tool in the controller design of high order nonlinear system [3]. The essence of back-stepping control is recursively design of appropriate state as pseudo control input for lower dimension subsystems. However, back-stepping control suffers from repeated differentiations of the virtual control inputs, so-called "explosion of complexity" [4]. As a result, the complexity of the controller grows drastically when the order of system dynamics is high. In [5, 6], differentiators are employed to construct the derivatives of the virtual control inputs and it solves the problem. In order to deal with unmatched disturbances in back-stepping design, the disturbance observer is introduced to improve the robustness in [7].

To achieve the adaptive design of integrated guidance and autopilot system, this paper combines back-stepping technique and L_1 adaptive control. L_1 adaptive control can adapt to varying system dynamics, cancel out uncertainties and nonlinearities. In [8], L_1 adaptive control architecture is introduced, which has guaranteed transient and steady performance bounds for system's input and output signals in the presence of fast adaptation without any gain scheduling in the controller parameters, and without high gain feedback. It is a powerful and low-cost controller in the presence of large uncertainties, un-modelled dynamics, and disturbances. The features have been verified in a large number of flight tests in recent papers [9-11]. In [9], output feedback L_1 adaptive control is designed and applied to missile guidance and control system. It demonstrates robustness to significant changes in the missile dynamics up to $\pm 50\%$ change in the missile longitude model parameters. In [10], the augmented base-line controller is used to improve angle of attack (AOA) control of an unstable aircraft with multiple control surfaces. In [11], it mainly addresses the application of L_1 adaptive control in IGC design for hypersonic glider. A dynamics pole placement controller is implemented as the baseline controller, and it is augmented with an L1 adaptive controller to cancel out the matched and unmatched uncertainties. In [12], control input constraint is considered in L1 adaptive controller with a simple form.

This paper is organized as follows: In Section 2, the nonlinear integrated guidance and autopilot model is built with uncertain aerodynamics coefficient parameters and time-varying disturbances, and a second order actuator dynamics is considered. In Section 3, the hybrid back-stepping controller is designed, first and third order nonlinear differentiators are used to avoid the direct differential of virtual control input. In Section 4, numerical simulations based on CAV-L are made to verify the proposed scheme. Two scenarios with aerodynamics coefficients uncertainties are set to track two kinds of FPA trajectories in hypersonic glide phase. In Section 5, the paper is concluded.

2. Problem Formation

In this section, the vertical flight dynamics is developed in the presence of uncertain aerodynamic parameters. The planar geometry of the hypersonic glider is depicted in figure 1. The unpowered glide phase can be divided into dive phase, pull-up phase and quasi-glide phase in sequence based on the characteristic of FPA trajectory. Thus different kinds of FPA trajectory are set in the numerical simulation. In figure 1, x_b is the symmetrical body axis. The coordinate systems are built without considering the earth rotation. γ, α, ω_z are respectively the FPA, AOA and pitch angle rate, δ_z is the control surface deflection. The angle and angle rate directions shown in figure 1 are positive. V is constant glide velocity, m, g are constant mass and gravity acceleration

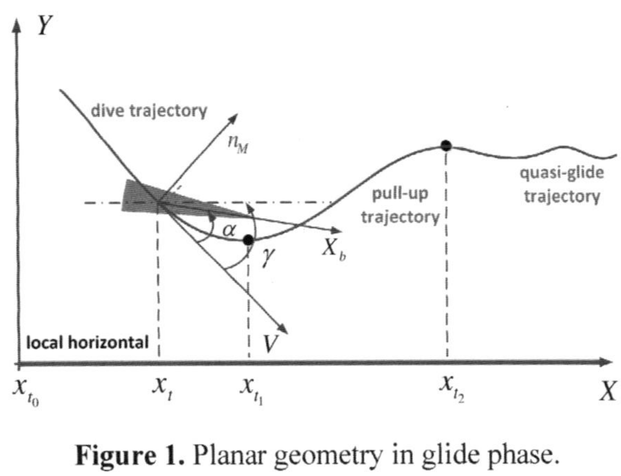

Figure 1. Planar geometry in glide phase.

2.1. Longitudinal Dynamics

The nonlinear system dynamics is built as equation (1).

$$\dot{\gamma} = \bar{L}_o + \bar{L}_\alpha \alpha - \frac{g}{V}\cos\gamma \tag{1a}$$

$$\dot{\alpha} = \omega_z - \bar{L}_o - \bar{L}_\alpha \alpha + \frac{g}{V}\cos\gamma \tag{1b}$$

$$\dot{\omega}_z = \bar{M}_o + \bar{M}_\delta \delta_z \tag{1c}$$

When the flight velocity is high, the lift force coefficient can be expressed as $C_L = C_L^\alpha \alpha + C_{Lo}$, so the lift consists of two parts:

$$L = \frac{1}{2} m\rho V^2 C_L = L_\alpha \alpha + L_o \tag{2a}$$

where C_L^α is the lift coefficient derivative with respect to AOA, and $L_o = \frac{1}{2} m\rho V^2 C_{Lo}$ is the other factors contributing to lift. The two parts of lift are unified by momentum as following:

$$\bar{L}_\alpha = \frac{L_\alpha}{mV} = \frac{1}{2}\rho V C_L^\alpha, \quad \bar{L}_o = \frac{L_o}{mV} = \frac{1}{2}\rho V C_{Lo} \tag{2b}$$

Also, the aerodynamics moment is described as $M_z = M_{\delta_z}\delta_z + M_o$, M_{δ_z} is the moment coefficient derivative with respect to δ_z, and M_o is the other factors contribution to moment. Both are unified by the moment of inertia with respect to z-axis I_z:

$$\bar{M}_o = \frac{M_o}{I_z}, \quad \bar{M}_{\delta_z} = \frac{M_{\delta_z}}{I_z} \tag{3}$$

2.2. Integrated Guidance and Autopilot

Redefining the flight sates as $x_1 = \gamma$, $x_2 = \alpha$, $x_3 = \omega_z$ and $x_4 = \delta_z$. Then equation (1) is reshaped as follows:

$$\begin{aligned} \dot{x}_1 &= f_1(x,t) + x_2 \\ \dot{x}_2 &= f_2(x,t) + x_3 \\ \dot{x}_3 &= f_3(x,t) + bx_4 \end{aligned} \tag{4}$$

The nonlinear functions with unmatched time-varying disturbances $d_i(t)$, $i = 1, 2, 3$ in equation (4) are depicted as follows:

$$f_1(x,t) = \bar{L}_o + (\bar{L}_\alpha - 1)x_2 - \frac{g}{V}\cos(x_1) + d_1(t) \tag{5a}$$

$$f_2(x,t) = -\bar{L}_o - \bar{L}_\alpha x_2 + \frac{g}{V}\cos(x_1) + d_2(t) \tag{5b}$$

$$f_3(x,t) = \bar{M}_o + d_3(t), \; b = \bar{M}_\delta \tag{5c}$$

We have two assumptions about the nonlinear functions. Firstly, there exists $B > 0$, such that $|f_i(t,0)| \le B$, $i = 1, 2, 3$, holds for all $t \ge 0$. Secondly, assuming that $f_i(x,t)$ is uniform bounded and the partial derivatives are semi-global uniform bounded. Then the following upper bound is yielded:

$$\|f_i(x,t)\|_{L_\infty} \le L_{\rho_r} \|x\|_{L_\infty} + B \tag{6}$$

where L_{ρ_r} is the bound of partial derivatives of the nonlinear functions $f_i(x,t)$. Besides $b \in [b_l, b_u]$, $0 < b_l < b_u$ are known conservative bounds.

For the hypersonic glider, the control objective is to design a robust controller in the presence of uncertainties and disturbances to ensure the system output $y = x_1 = \gamma$ to track the reference signal under $x_4 = \delta_z$. Additionally, the dynamics of actuator is integrated with the vertical flight. This paper considers a second order actuator dynamics given by the following transfer function:

$$G(s)=\frac{\delta_z(s)}{\delta_{zc}(s)}=\frac{K_{\delta_z}}{s^2-2\omega_n\xi s+\omega_n^2} \tag{7}$$

By defining $x_5=\dot{\delta}_z$, $u=\delta_{zc}$, the sate space model of actuator dynamics is built as follows with uncertainty and disturbance.

$$\begin{aligned}\dot{x}_4 &= x_5\\ \dot{x}_5 &= f_r(x_4,x_5)+\Delta f_r+b_r u(t)+d_r(t)\end{aligned} \tag{8}$$

where

$$f_r(x_4,x_5)=2\omega_n\xi x_5-\omega_n^2 x_4,\ b_r=K_{\delta_z} \tag{9}$$

In equation (8), Δf_r is the uncertainty and $d_r(t)$ is the time-varying disturbance. They are bounded under positive constant F, D .

$$|\Delta f_r|\le F,\ |d_r|\le D \tag{10}$$

To get a strict feedback form of the state space, RBF neural network (RBFNN) is implemented to make $f_1(x,t)$ only relate to (x_1,t). RBFNN is one of the most widely used networks models in nonlinear control systems. Theoretically, as long as a sufficient number of neurons are employed, RBFNN can approximate any continuous function to an arbitrary accuracy on any compact set.

$$f_1(x_1)=\Phi(x_1)-\frac{g}{V}\cos(x_1)+d_1(t) \tag{11}$$

In equation (11), $\Phi(\cdot)$ is depicted as follows:

$$\Phi(x_1)=\theta^T\xi(x_1)+\Delta \tag{12}$$

where Δ is neural network functional approximation error, x_1 is the input of RBFNN, $\theta=[\theta_1,\ \theta_2\ \cdots\ \theta_n]^T\in R^n$ is the weight vector to be determined, and n is the node number of RBFNN. $\xi(x_1)$ is a vector of Gaussian functions of equation (13).

$$\xi_j(x_1)=\exp\left(-\frac{\|x_1-\varepsilon_j\|_2^2}{2\beta^2}\right),\ j=1,\ 2,\ 3....\ n \tag{13}$$

$\varepsilon=[\varepsilon_1,\ \varepsilon_2,\ \cdots\ \varepsilon_n]^T$ is the central vector of the Gaussian function, and β is the width. In order to realize small reconstruction errors, it is necessary to use high-order networks. The weighing vector updates as the following adaptive law:

$$\dot{\theta}=-\lambda(e_{rbf}\xi+\sigma\theta) \tag{14}$$

Where λ, σ is positive constant, e_{rbf} is the estimate error. Assuming there has an ideal estimate of $\Phi_d(x_1)$, then the optimal weighing vector is given as follows:

$$\theta^*=\arg\min\left[\sup|\Phi_d-\Phi|\right] \tag{15}$$

3. Controller Design

In this section, back-stepping L_1 adaptive controller is designed for IGC system depicted as equation (5). This state feedback controller mainly compensates the uncertainties of the nonlinear plant with the help of low-pass command filters $C(s)$. The filters ensure the controller remains in low frequency range in the presence of fast adaptation and large reference input. Here this adaptive control scheme is chosen for the following reasons: 1) fast adaption without high gain; 2) great robustness to large system uncertainties and different reference signals; 3) reduced cost for the validation and verification process. Then sliding mode controller with nonlinear differentiator is implemented for the autopilot to track the control surface deflection.

3.1. Command Filters

Let

$$\boldsymbol{C}(s)=\begin{bmatrix} C_1(s) & & \\ & C_2(s) & \\ & & C_3(s) \end{bmatrix} \tag{16}$$

In this case, the design of L_1 adaptive controller involves three stable low-pass filters. Technically, any transfer function can be used as long as the filters keep the stability of the state feedback controller. Simple first and second order linear transfer function can often result in good performance. Here $C_1(s)$, $C_2(s)$ is verified by relative degree of at least 2 within unity DC gain, i.e. $C_1(0)=1$, $C_2(0)=1$. $C_2(s)$, $C_3(s)$ have positive feedback gains k_2, k_3 and strictly proper transfer functions $D_2(s)$, $D_3(s)$.

$$C_1(s)=\frac{k_1D_1(s)}{1+k_1D_1(s)},\ C_2(s)=\frac{k_2D_2(s)}{1+k_2D_2(s)},\ C_3(s)=\frac{bk_3D_3(s)}{1+bk_3D_3(s)} \tag{17}$$

$C_3(s)$ is also within unity DC gain.

3.2. State Predictor

Under the assumptions of the nonlinear functions, the IGC model can be approximated as follows:

$$\begin{aligned} \dot{x}_1 &= \theta_1\|x\|_\infty+\sigma_1+x_2 \\ \dot{x}_2 &= \theta_2\|x\|_\infty+\sigma_2+x_3 \\ \dot{x}_3 &= \theta_3\|x\|_\infty+\sigma_3+bx_4 \end{aligned} \tag{18}$$

then consider the following state predictor:

$$\begin{aligned} \dot{\hat{x}}_1 &= -a_1\tilde{x}_1+\hat{\theta}_1\|x\|_\infty+\hat{\sigma}_1+x_2 \\ \dot{\hat{x}}_2 &= -a_2\tilde{x}_2+\hat{\theta}_2\|x\|_\infty+\hat{\sigma}_2+x_3 \\ \dot{\hat{x}}_3 &= -a_3\tilde{x}_3+\hat{\theta}_3\|x\|_\infty+\hat{\sigma}_3+\hat{b}u \end{aligned} \tag{19}$$

where $\hat{x}_1$, $\hat{x}_2$, $\hat{x}_3$ are the predictor states, $\tilde{x}_1=\hat{x}_1-x_1$, $\tilde{x}_2=\hat{x}_2-x_2$, $\tilde{x}_3=\hat{x}_3-x_3$ are the prediction error, $\hat{\theta}_1$, $\hat{\theta}_2$, $\hat{\theta}_3$, $\hat{\sigma}_1$, $\hat{\sigma}_2$, $\hat{\sigma}_3$, $\hat{b}$ are the adaptive estimates.

3.3. Adaptation

The adaptive estimates are governed by

$$\begin{aligned} &\dot{\hat{\theta}}_1=\Gamma\,\mathrm{Proj}\left(\hat{\theta}_1,-\tilde{x}^T\boldsymbol{P}[1,0,0]^T\|x\|_\infty\right),\ \dot{\hat{\sigma}}_1=\Gamma\,\mathrm{Proj}\left(\hat{\sigma}_1,-\tilde{x}^T\boldsymbol{P}[1,0,0]^T\right) \\ &\dot{\hat{\theta}}_2=\Gamma\,\mathrm{Proj}\left(\hat{\theta}_2,-\tilde{x}^T\boldsymbol{P}[0,1,0]^T\|x\|_\infty\right),\ \dot{\hat{\sigma}}_2=\Gamma\,\mathrm{Proj}\left(\hat{\sigma}_2,-\tilde{x}^T\boldsymbol{P}[0,1,0]^T\right) \\ &\dot{\hat{\theta}}_3=\Gamma\,\mathrm{Proj}\left(\hat{\theta}_3,-\tilde{x}^T\boldsymbol{P}[0,0,1]^T\|x\|_\infty\right),\ \dot{\hat{\sigma}}_3=\Gamma\,\mathrm{Proj}\left(\hat{\sigma}_3,-\tilde{x}^T\boldsymbol{P}[0,0,1]^T\right) \\ &\dot{\hat{b}}=\Gamma\,\mathrm{Proj}\left(\hat{b},-\tilde{x}^T\boldsymbol{P}[0,0,1]^T u\right) \end{aligned} \tag{20}$$

Let $\boldsymbol{A}_m$ be defined as:

$$\boldsymbol{A}_m=diag\left[-a_1\quad -a_2\quad -a_3\right] \tag{21}$$

where a_1, a_2, $a_3>0$ are positive constants specifying the desired closed-loop dynamics. $\Gamma\in\mathrm{R}^+$ is the adaptation gain, and the symmetric positive definite matrix $\boldsymbol{P}=\boldsymbol{P}^T>0$ solves the Lyapunov equation $\boldsymbol{A}_m^T\boldsymbol{P}+\boldsymbol{P}\boldsymbol{A}_m=-\boldsymbol{Q}$ for arbitrary $\boldsymbol{Q}=\boldsymbol{Q}^T>0$. The projection operator $\mathrm{Proj}(\bullet,\bullet)$ ensures that $\left|\hat{\theta}_i(t)\right|\le\theta_b$, $\hat{\sigma}_i\le\Delta$, $i=1,\ 2,\ 3$ and $\hat{b}(t)\in\Omega$.

Let

$$\boldsymbol{A}_g = \begin{bmatrix} -a_1 & 1 & \\ & -a_2 & 1 \\ & & -a_3 \end{bmatrix} \tag{22}$$

and $G(s) \triangleq H(s)(\boldsymbol{I} - \boldsymbol{C}(s))$, $H(s) \triangleq (s\boldsymbol{I} - \boldsymbol{A}_g)^{-1}$, the selection of $\boldsymbol{C}(s)$ need to ensure the following L_1 –norm upper bound can be verified.

$$\|G(s)\|_{L_1} < \frac{\rho_r - P}{K_1 \rho_r + K_2} \tag{23}$$

where K_1, K_2, P are constants related to ρ_r and B.

3.4. Control law

The adaptive controller for the integrated guidance and autopilot system made up by equation (5) and (8) is designed step-by-step. Consider the first subsystem:

$$\dot{x}_1 = \theta_1 \|x\|_\infty + \sigma_1 + \overline{x}_2 \tag{24}$$

In equation (24), $\overline{x}_2$ is the virtual control input, it's depicted as follows:

$$\overline{x}_2 = -a_1(x_1 - r) - \hat{\eta}_{1C}(t) + \dot{r} \tag{25a}$$

$$\hat{\eta}_{1C}(s) = C_1(s)\hat{\eta}_1(s) \tag{25b}$$

where

$$\hat{\eta}_1(t) = \hat{\theta}_1(t)\|x(t)\|_\infty + \hat{\sigma}_1 \tag{26}$$

$\hat{\eta}_1(s)$ is the Laplace transform of $\hat{\eta}_1(t)$.

Consider the second subsystem:

$$\dot{x}_2 = \theta_2 \|x\|_\infty + \sigma_2 + \overline{x}_3 \tag{27}$$

where

$$\overline{x}_3 = -k_2 D_2(s)\hat{\eta}_\alpha(s) \tag{28}$$

In equation (28), $\hat{\eta}_\alpha(s)$ is the Laplace transform of the following equation (29a).

$$\hat{\eta}_\alpha(t) = \hat{\eta}_2(t) + a_2(x_2(t) - \overline{x}_2(t)) - \dot{\overline{x}}_2(t) \tag{29a}$$

$$\hat{\eta}_2(t) = \overline{x}_3(t) + \hat{\theta}_2(t)\|x(t)\|_\infty + \hat{\sigma}_2(t) \tag{29b}$$

The third subsystem is depicted as

$$\dot{x}_3 = \theta_3 \|x\|_\infty + \sigma_3 + b\overline{x}_4 \tag{30}$$

The control input is depicted as follows:

$$\overline{x}_4(s) = -k_3 D_3(s)\hat{\eta}_u(s) \tag{31}$$

where $\hat{\eta}_u(s)$ is the Laplace transform of the signal in equation (32a).

$$\hat{\eta}_u(t) = \hat{\eta}_3(t) + a_3(x_3(t) - \overline{x}_3(t)) - \dot{\overline{x}}_3(t) \tag{32a}$$

$$\hat{\eta}_3(t) = \hat{b}(t)u(t) + \hat{\theta}_3(t)\|x(t)\|_\infty + \hat{\sigma}_3(t) \tag{32b}$$

To get the second order derivatives of $\overline{x}_4$, a third order differentiator is implemented:

$$\begin{aligned} \dot{x}_{thod1} &= x_{thod2} \\ \dot{x}_{thod2} &= x_{thod3} \\ \dot{x}_{thod3} &= \frac{4}{\varepsilon^3}\left(-2^{\frac{5}{3}}\left(|v|^{\frac{1}{3}}\operatorname{sgn}(v)\right) - \left|\varepsilon^2 x_{thod3}\right|^{\frac{5}{3}}\operatorname{sgn}\left(\varepsilon^2 x_{thod3}\right)\right) \\ v &= x_{thod1} - x_{4u} + \left|\frac{1}{\varepsilon}x_{thod2}\right|^{\frac{9}{7}}\operatorname{sgn}(\varepsilon x_{thod2}),\ x_{thod1}(0) = \overline{x}_4(0) \end{aligned} \tag{33}$$

The error between x_{thod1} and $\bar{x}_4$ can be enough small through choosing suitable ε. Then x_{thod2} and x_{thod3} are respectively the first and second order derivatives of $\bar{x}_4$.

Defining a scalar $e_r = x_4 - \bar{x}_4$ and a vector $E = [e_r,\ \dot{e}_r]^T$, then the following nonlinear sliding mode surface is designed.

$$\sigma = CE - CP \tag{34}$$

In equation (34), $C = [c_1,\ c_2]$ is a constant vector, and $P = [p,\ \dot{p}]^T$ is determined by the following nonlinear function:

$$p(t) = \begin{cases} e_r + \dot{e}_r t + \frac{1}{2}\ddot{e}_r t^2 - \left(\frac{10}{T^3}e_r + \frac{6}{T^2}\dot{e}_r + \frac{3}{2T}\ddot{e}_r\right)t^3 + \left(\frac{15}{T^4}e_r + \frac{8}{T^3}\dot{e}_r + \frac{3}{2T^2}\ddot{e}_r\right)t^4 \\ -\left(\frac{6}{T^5}e_r + \frac{3}{T^4}\dot{e}_r + \frac{1}{2T^3}\ddot{e}_r\right)t^5,\ 0 \le t \le T \\ 0,\ t > T \end{cases} \tag{35}$$

It can be seen that T is the convergence time of this nonlinear sliding mode controller. Finally, the control input of integrated guidance and autopilot is designed as follows:

$$u = -\frac{1}{b_r}\left(f_r(x) - x_{thod3} - \ddot{p} + \frac{c_1}{c_2}(\dot{e}_r - \dot{p})\right) - \frac{1}{b_r}\operatorname{sgn}(c_2\sigma)(F + D) \tag{36}$$

4. Numerical Simulation

To verify numerically the performance of the proposed method, two simulation scenarios with aerodynamics force and moment coefficient uncertainties are performed. The aerodynamics coefficients are chosen based on CAV-L in Appendix A. Two kinds of FPA trajectory are set as follows:

$$r(t) = 10\sin\left(\frac{2\pi}{15}t\right) + 3 \tag{37a}$$

$$r(t) = \begin{cases} 2t\ ,t \ge 15 \\ -2t + 60\ ,t < 15 \end{cases} \tag{37b}$$

Scenario 1:

$$\begin{aligned} &\bar{L}_o = -1.25,\ \bar{L}_\alpha = 1.1 \\ &f_1(x_1,t) = \Phi(x_1) - 0.1\cos(x_1) - 0.2\sin(0.1t) \\ &f_2(x_1,x_2,t) = -0.25 - 1.1x_2 + 0.1\cos(x_1) - 0.2\sin(0.1t) \\ &f_3(x,t) = 1 \\ &b = 0.8 \end{aligned} \tag{38}$$

Scenario 2:

$$\begin{aligned} &\bar{L}_o = -1.25,\ \bar{L}_\alpha = 1.1 + 0.1x_2 \\ &f_1(x_1,t) = -1.25 + \Phi(x_1) - 0.1\cos(x_1) - 0.2\sin(0.1t) \\ &f_2(x_1,x_2,t) = -1.25 - 0.1x_2^2 - 1.1x_2 + 0.1\cos(x_1) - 0.2\sin(0.1t) \\ &f_3(x,t) = 0.01x_1 + 0.01x_2 + 0.01x_3 + 0.05\sin(t) \\ &b = 0.8 \end{aligned} \tag{39}$$

Comparing with Scenario 1, the lift force coefficient in Scenario 2 has factor of AOA and the moment is related to flight states. The second order actuator dynamics is set with $\omega_n = 30,\ \xi = 0.3$, and is saturated with magnitude maximum 30deg. However, a single L_1 is designed for the two Scenarios with the following control parameters:

The state predictor is given by $a_1 = a_2 = a_3 = 2$. The adaptation is determined by $\boldsymbol{Q} = diag[150\ \ 1\ \ 1]$, and the projection bound is set as $\Omega = [0.1,\ 3]$, the adaptation gain is $\Gamma = 10000$.

The three filters are set as follows:

$$C_1(s)=\frac{3}{s^2+1.1s+1.3} \tag{40a}$$

$$C_2(s)=\frac{100}{s^2+15s+100} \tag{40b}$$

$$C_3(s)=\frac{30}{s+30} \tag{40c}$$

The RBFNN is setting as follows:

$$j=9,\ b=2.5,\ \lambda=0.15,\ \sigma=0.5,\ \varepsilon=[-1\ \ -0.7\ \ -0.5\ \ -0.3\ \ 0\ \ 0.3\ \ 0.5\ \ 0.7\ \ 1].$$

Here the estimate error of RBFNN can be combined in the system uncertainties.

The simulation results of tracking sinusoidal FPA reference trajectory in equation (37a) are shown in figure 2 to figure 5 and that of step FPA reference trajectory in equation (37b) are shown in figure 6 to figure 9. From figure 2 and figure 6, it can be seen that the proposed IGC control method can track the reference FPA trajectory under lift and moment coefficient uncertainties and time-varying disturbances in different channel. Figure 3 shows the AOA curves under reference FPA of equation (37a), and compares the AOA curve of Scenario 1 with Scenario 2. The second order item of AOA in S2 makes the curve smoother. Figure 4 shows the pitch angle rate comparison. The control surface deflections are presented in figure 5, which shows that the nonlinearity of moment coefficients results in control surface deflection chasing in a small range.

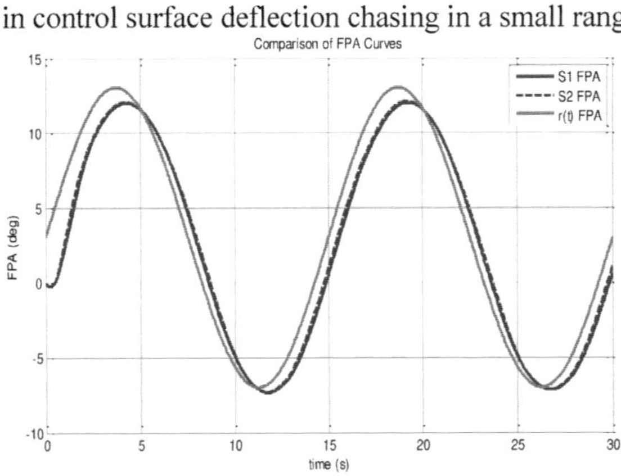

Figure 2. The tracking result of FPA

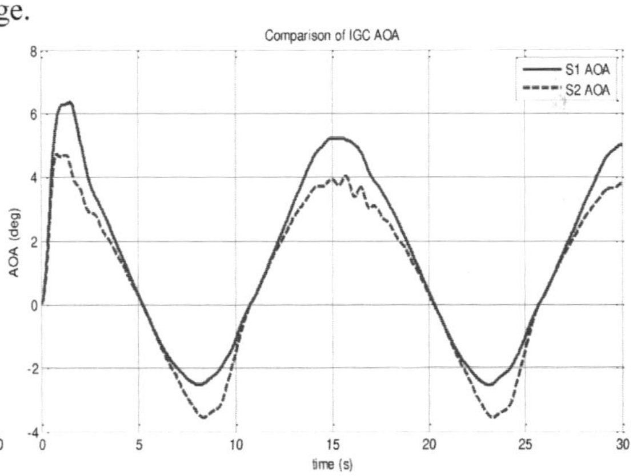

Figure 3. The AOA comparison S1 with S2

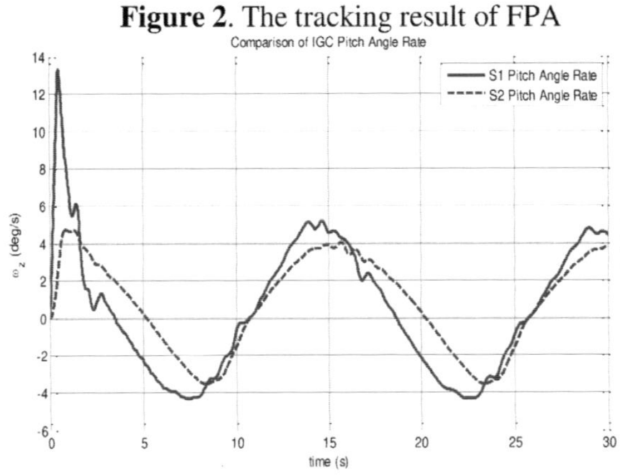

Figure 4. The pitch angle rate comparison

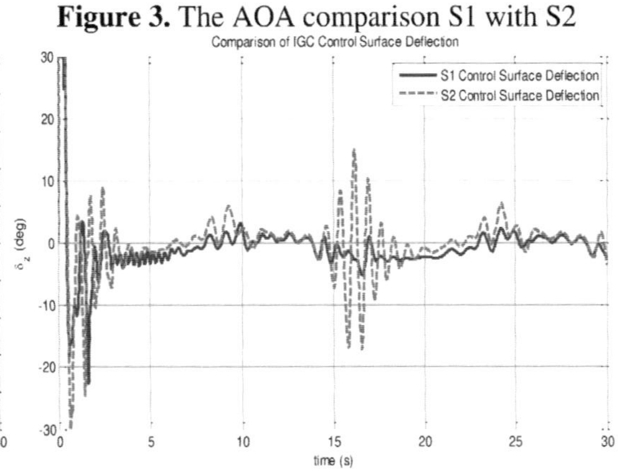

Figure 5. The control surface deflection comparison

Figure 7 shows the AOA curves under reference FPA of equation (37b), and compares the AOA curve of Scenario 1 with Scenario 2. Figure 8 shows the pitch angle rate comparison. The control surface deflections are presented in Figure 9.

IOP Conf. Series: Earth and Environmental Science **104** (2017) 012010 doi:10.1088/1755-1315/104/1/012010

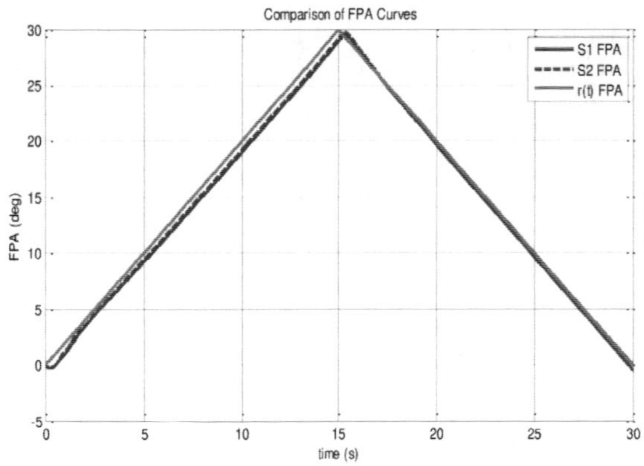

Figure 6. The tracking result of FPA

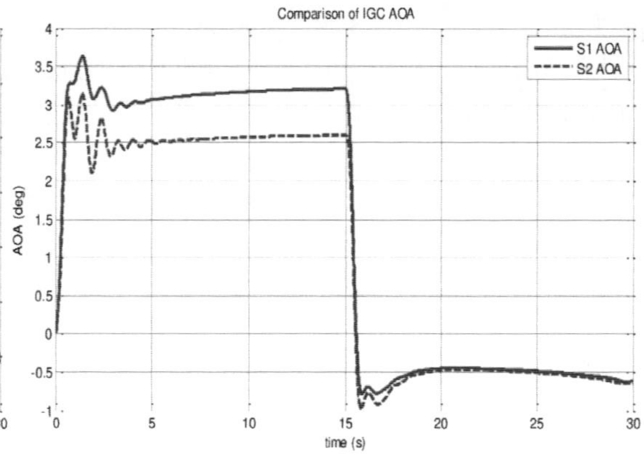

Figure 7. The AOA comparison S1 with S2

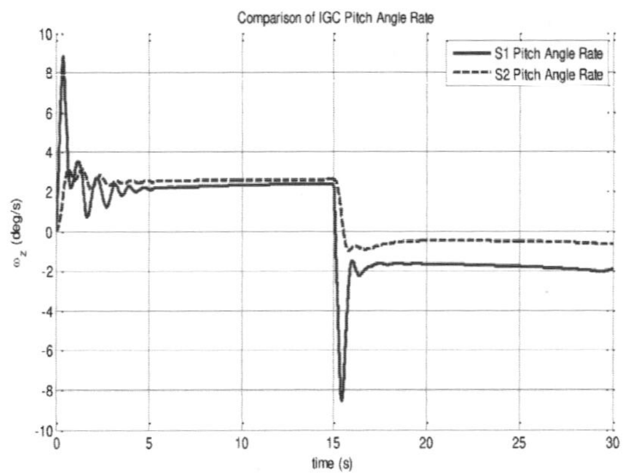

Figure 8. The pitch angle rate comparison

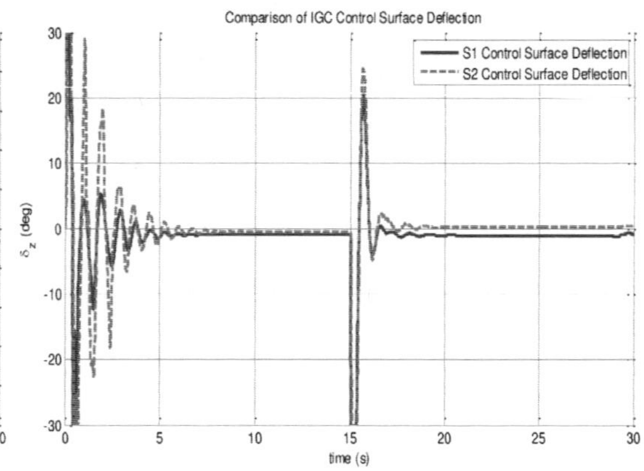

Figure 9. The control surface deflection comparison

5. Conclusion

Motivated by the reviews and discussions on the integrated guidance and control design, adaptive command filtered control is firstly proposed for the nonlinear IGC with second order actuator dynamics in the presence of aerodynamics coefficient uncertainties and disturbances for hypersonic glider in the glide phase. Through the numerical simulations of FPA tracking in different scenarios, it can be concluded that adaptive command filtered IGC can keep fast transient response and good robustness without changing the controller setting, thus significantly reduce the amount of Monte-Carlo analysis for V&V. What's more, the proposed IGC avoids input peaking and chasing, which means better in practical application.

6. Appendix A

The aerodynamics of NASA CAV-L in 1998 is performed in the following tables.

Table A.1. CAV-L Lift Coefficient

Mach / AOA	3.5	5	8	15	20
10°	0.3401	0.3264	0.3108	0.2856	0.2760
15°	0.5786	0.5358	0.4883	0.4991	0.4349
20°	0.7975	0.7291	0.6731	0.6137	0.5975

In hypersonic flight, the lift coefficient can be simplified as $C_L = C_L^\alpha \cdot \alpha + C_{L0}$. The fit results of lift coefficient in Table A.1 are shown in Figure A.1, and the parameters are listed in Table A.2.

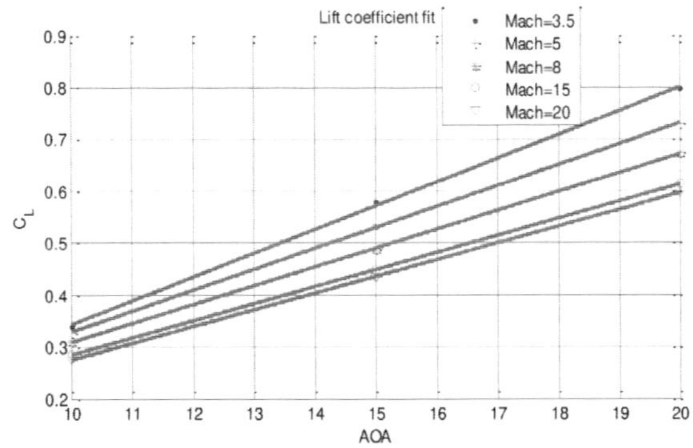

Figure A.1 Polynomial fit results

Take $Mach = 4,\ H = 20km$ and $S_{ref} = 0.35m^2$ as example to calculate the aerodynamics parameters:

$$\bar{L}_o = \frac{L_o}{mV} = \frac{1}{2}\rho V S_{ref} C_{Lo} = -1.673,\quad \bar{L}_\alpha = \frac{L_\alpha}{mV} = \frac{1}{2}\rho V S_{ref} C_L^\alpha = 0.9149$$

Table A.2. Fit curve parameters

Mach	C_L^α	C_{L0}
3.5	0.04574	-0.114
5	0.04027	-0.07362
8	0.03623	-0.05272
15	0.03281	-0.04268
20	0.03215	-0.4612

Acknowledgments
The authors would like to thank the reviewers and the editor for their comments and suggestions that helped to improve the paper significantly.

References

[1] Tal Shima, Moshe Idanm, Ode M. Colan, Sliding mode control for integrated missile autopilot and guidance, Journal of Guidance, Control, and Dynamics, 29(2): 250-260, 2006.
[2] Nathan Harl, S.N. Balakrishman, Craig Phillips, Sliding mode integrated missile guidance and control, AIAA Guidance, Navigation, and Control Conference, Toronto, Ontario Canada, 2010.
[3] Jay Farrell, Marios Polycarpou, Manu Sharma, Adaptive back-stepping with magnitude, rate, and bandwidth constraints: aircraft longitude control, Air Force Research Laborary Report, 2006.
[4] Hongli Shi, A novel scheme for the design of back-stepping control for a class of nonlinear systems, Applied Mathematical Modeling, 35(2011):1893-1903, 2010.
[5] Mingzhe Hou, Guangren Duan, Adaptive dynamic surface control for integrated missile guidance and autopilot, International Journal of Automation and Computing, 8(1): 122-127, 2011.
[6] Xiaoling Liang, Mingzhe Hou, and Guangren Duan, Adaptive dynamic surface control for integrated missile guidance and autopilot in the presence of input saturation. Journal of Aerospace Engineering, 0414121-1-8, 2014.

[7] Dong Chaoyang, Cheng Haoyu, Wang Qing, Back-stepping sliding mode control for integrated guidance and control design based on active disturbance rejection. Systems Engineering and Electronics, 37(7): 1604-1609, 2015.

[8] Cao, C. and Hovakimyan, N., Design and analysis of a novel L_1 adaptive control architecture with guaranteed transient performance, IEEE Transactions on Automatic Control, 53(2): 586-591, 2008.

[9] Jiang Wang, Chengyu Cao, L_1 adaptive control for a missile longitudinal autopilot, AIAA Guidance, Navigation and Control Conference and Exhibit, 2008.

[10] Ola Harkegard, Modifying L_1 adaptive control for augmented angle of attack control, AIAA Guidance, Navigation, and Control Conference, 2009.

[11] Sanchito Banerjee, L_1 adaptive control augmentation for the longitudinal dynamics of a hypersonic glider, Journal of Guidance, Control, and Dynamics, 39(2): 275–291, 2016.

[12] Wu Wenhai, Gao Li, Mei Dan, Zhou Siyu, L_1 adaptive controller for aircraft attitude with input constraint, Journal of Nanjing University of Aeronautics & Astronautics, 44(6): 809-816, 2012.

Analysis of Factors Influencing $PM_{2.5}$ in Beijing: A Microcosmic and Dynamic Perspective for Sustainable Development

Yani Wang[1], Jun Wang[1,*] and Guiping Tao[2]

[1]School of Economics and Management, Beihang University, Xueyuan Road 37, Haidian District, Beijing 100191, PR China
[2]School of statistics, Capital University of Economics and Business, Huaxiang Road 121, Fengtai District, Beijing 100070, PR China

*king.wang@buaa.edu.cn

Abstract. Haze pollution has become a hot issue concerned with the process of modernization and one serious problem requiring urgent solution, especially in Beijing. $PM_{2.5}$ is the main reason causing haze and its harm. Although there has been research centering on factors affecting $PM_{2.5}$, little attention has been devoted to the microcosmic and dynamic effects on it. Vector auto-regression (VAR) mode is applied in this study to explore the interaction between $PM_{2.5}$, PM_{10}, SO_2, CO and NO_2. Results of Granger causality tests tell that there exists causal relationship between PM_{10}, SO_2, CO, NO_2 and $PM_{2.5}$. Impulse response functions (IRFs) show that the response of $PM_{2.5}$ to a shock in CO is positive and large in the short period, while the reaction of $PM_{2.5}$ to a shock in SO_2 increases over time. Meanwhile, variance decomposition indicate that $PM_{2.5}$ is more closely related to CO in the short term while SO_2' influence accounts for a higher proportion in the long run. The findings provide a novel perspective to analyze the factors influencing $PM_{2.5}$ dynamically and contribute to a better understanding of haze and its relationship with sustainable development.

1. Introduction

Nowadays, air pollution has been a big problem for people's normal work, life and health, from which Beijing has a narrow escape. Air pollution in Beijing may be more serious than that in other cities, and haze weather in recent years is a critical factor causing severe environmental crisis and has caused widespread concern [1].

The biggest source of haze in the air is what is called PM (Particulate Matter) particles, which are divided into PM_{10} and $PM_{2.5}$ according to their diameter sizes [2]. The smaller the particles are, the more tenacious their vitality is and they can do more harm to human beings. Furthermore, they can invade the body with people's breathing and cause harm to the human body. In 2012, particles whose diameter is less than 2.5 micrometers were collectively referred to as 'the fine particulate matter' by the U.S. environmental protection agency, also known as $PM_{2.5}$ [3]. The average width of the fine particles reaches about one of 30 points of human hair. Because of their extremely small size, they can be deep into the lungs and therefore do much harm to human health. Ten years' studies have shown long-term exposure to $PM_{2.5}$ can cause cardiovascular mortality rate increased by 6% -13% [4].

With haze increasingly rampant, the research and governance on haze has become a global work. Cheng et al. studied characteristic of particulate matter in Alberta cities, founding that the contribution of various pollutants is different, but on the whole vehicle emissions and fuel combustion play important roles [5]. Sawant et al. specifically analyzed the chemical composition of $PM_{2.5}$ in Southern California, with results showing that the region's concentration of elemental carbon OC is much higher than that in other cities, which is critical to the region's high concentration of $PM_{2.5}$ [6]. Matthew's research about $PM_{2.5}$ components indicated that sulfur compounds are the main components of $PM_{2.5}$ and the sulphide is largely decided by the concentration of SO_2 [7]. David et al. (2013) identified the major sources of $PM_{2.5}$ and their impact on environmental and human health [1]. Cox analyzed $PM_{2.5}$ and Air Pollutants using regression analysis [3], while there is research assessing the health and economic value of the smog pollution, lacking causes and prevention measures [8].

Previous studies on $PM_{2.5}$ are mainly limited to macro level. Our studies offer an important original contribution to examining the factors affecting $PM_{2.5}$ from the microcosmic and dynamic perspective. Vector auto-regression model combing with IRFs and variance decomposition is established to explore the interaction between PM_{10}, CO, SO_2, NO_2 and $PM_{2.5}$, hoping to provide theoretical foundation and experience for future governance of haze in Beijing.

2. Empirical Analysis

2.1. Data

China has started to implement new air quality index (AQI) since January 2014, and $PM_{2.5}$ is included in the calculation of AQI [9]. Indicators are concentration of $PM_{2.5}$, PM_{10}, CO, SO_2 and NO_2 and corresponding data is extracted from the AQI monitoring system and covers the period of January 1, 2014 to December 31, 2014. Correlation coefficient matrix is calculated to indicate the correlation among these indicators and we find that correlation coefficients are all above 0.5, some even are close to 0.9, indicating there exists positive and significant correlations between $PM_{2.5}$, PM_{10}, CO, SO_2 and NO_2. Consequently, it is necessary to analyze the interplay among them.

2.2. Empirical Methodology

We begin our empirical analysis by VAR mode to estimate the interactions between $PM_{2.5}$, PM_{10}, CO, SO_2 and NO_2. We then address several econometric issues that might arise in this study, namely Granger causality tests, Impulse Response Functions and Variance decomposition.

The advantages of VAR approach to traditional regression are in the following: First, it treats all variables as joint endogenous without any prior constraints, exploring the bidirectional causality between all variables. Secondly, the mode takes lagged effects into consideration, each endogenous variable can be the function of all variables lag item. Third, it can illustrate the effects of a shock in one variable on others applying IRFs and variance decomposition [10].

The general form of the VAR model is as follows:

$$\begin{bmatrix} PM_{2.5_t} \\ PM_{10_t} \\ CO_t \\ SO_{2_t} \\ NO_{2_t} \end{bmatrix} = \begin{bmatrix} C_{PM_{2.5}} \\ C_{PM_{10}} \\ C_{CO} \\ C_{SO_2} \\ C_{NO_2} \end{bmatrix} + \sum_{q=1}^{Q} \begin{bmatrix} a_{11}^{t-q} & a_{12}^{t-q} & a_{13}^{t-q} & a_{14}^{t-q} & a_{15}^{t-q} \\ a_{21}^{t-q} & a_{22}^{t-q} & a_{23}^{t-q} & a_{24}^{t-q} & a_{25}^{t-q} \\ a_{31}^{t-q} & a_{32}^{t-q} & a_{33}^{t-q} & a_{34}^{t-q} & a_{35}^{t-q} \\ a_{41}^{t-q} & a_{42}^{t-q} & a_{43}^{t-q} & a_{44}^{t-q} & a_{45}^{t-q} \\ a_{51}^{t-q} & a_{52}^{t-q} & a_{53}^{t-q} & a_{54}^{t-q} & a_{55}^{t-q} \end{bmatrix} \begin{bmatrix} PM_{2.5_{t-q}} \\ PM_{10_{t-q}} \\ CO_{t-q} \\ SO_{2_{t-q}} \\ NO_{2_{t-q}} \end{bmatrix} + \begin{bmatrix} \varepsilon_{PM_{2.5_t}} \\ \varepsilon_{PM_{10_t}} \\ \varepsilon_{CO_t} \\ \varepsilon_{SO_{2_t}} \\ \varepsilon_{NO_{2_t}} \end{bmatrix} \quad (1)$$

Where $PM_{2.5t}$ denotes the concentration of $PM_{2.5}$ in t. Q represents the order of the model, which can be determined by using Akaik' s information criterion (AIC). We calculate AIC for each model and determine the optimal lag length 2 according to the principle of the minimum AIC [11]. Next, Granger causality test is used to analyze the causal relationship between pairs of variables, such as testing whether CO is the cause of $PM_{2.5}$.

The VAR model is also supplemented with IRFs and Variance decomposition to elaborate the dynamic relationship between variables. IRFs are used to examine the response of one variable to a shock in another and check whether the impact is transitory or longer term. Variance decomposition explains the variation of variance on one variable caused by other variables. Specifically, it can distinguish the importance of all variables by analyzing their contribution to a certain variable concerning the change of its variance. Overall, IRFs and variance decomposition give us a better understanding of how PM_{10}, SO_2, CO and NO_2 influence $PM_{2.5}$ from the microcosmic and dynamic perspective.

3. Results

To conduct both Granger causality and VAR analysis, the variables must be stationary. Results of the test indicate that all variables are stationary. Granger causality tests also indicate that the joint variables have causal relationship. This supports our approach of analyzing the variables as a full dynamic system through VAR analysis.

3.1. VAR Regression Results

VAR model is conduct to study the interaction between SO_2, CO, NO_2, PM_{10} and $PM_{2.5}$. The VAR equation is as follows:

$$\begin{bmatrix} PM_{2.5} \\ PM_{10} \\ SO_2 \\ CO \\ NO_2 \end{bmatrix} = \begin{bmatrix} 29.28 \\ 38.94 \\ 49.63 \\ 9.95 \\ 23.51 \end{bmatrix} + \begin{bmatrix} 0.85 & 0.09 & 0.10 & 0.03 & 0.99 \\ 0.28 & 0.32 & 0.41 & 0.04 & 1.41 \\ 0.06 & -0.05 & 0.81 & -0.01 & 0.23 \\ 2.30 & -2.03 & 7.83 & 0.08 & 14.05 \\ 0.02 & -0.05 & 0.17 & -0.01 & 0.96 \end{bmatrix} \begin{bmatrix} PM_{2.5}(-1) \\ PM_{10}(-1) \\ SO_2(-1) \\ CO(-1) \\ NO_2(-1) \end{bmatrix} + \begin{bmatrix} 0.15 & -0.12 & 0.76 & 0.01 & -0.01 \\ 0.09 & -0.1 & 0.63 & 0.01 & 0.20 \\ -0.02 & -0.01 & -0.21 & 0.00 & 0.20 \\ 0.97 & 0.92 & 6.31 & 0.01 & 1.22 \\ -0.01 & -0.01 & 0.15 & 0.00 & 0.09 \end{bmatrix} \begin{bmatrix} PM_{2.5}(-2) \\ PM_{10}(-2) \\ SO_2(-2) \\ CO(-2) \\ NO_2(-2) \end{bmatrix} \quad (2)$$

We can see that $PM_{2.5}$ has positive relationship with itself in short-term supported by the positive and significant coefficient 0.85 on the first lag. CO has a short-term positive relationship with $PM_{2.5}$, which is shown by the positive and significant coefficient on the first lag and insignificant coefficient for the second lag. On the first lag, the relationship between SO_2 and $PM_{2.5}$ is insignificant. However, the coefficient is significant as time goes by and reaches 0.76 in the second lag. Besides, the influence of NO_2 on $PM_{2.5}$ declines with time going by. In addition, $PM_{2.5}$' influence on PM_{10} is positive and significant. Meanwhile, PM_{10} has positive relationship with itself on the first lag. Just as the influence imposing on $PM_{2.5}$, CO has a short-term positive relationship and the influence of SO_2 on PM_{10} can be manifested until the second lag. Besides, the influence of NO_2 on PM_{10} is significant on the first lag and declines with time going by. When it comes to the results with CO, SO_2 and NO_2 as dependent variables, we find that none of them are significantly influenced by $PM_{2.5}$ or PM_{10}.

3.2. Impulse Response Functions

We supplement the regression estimates with the analysis of IRFs. Although each variable in the system can be used as a response variable to the impact of the other variables, we focus on $PM_{2.5}$'s response to other variables. IRFs in our study can be understood as: the response of $PM_{2.5}$ to a shock in PM_{10}, SO_2, CO and NO_2 respectively. The IRFs results are as follows:

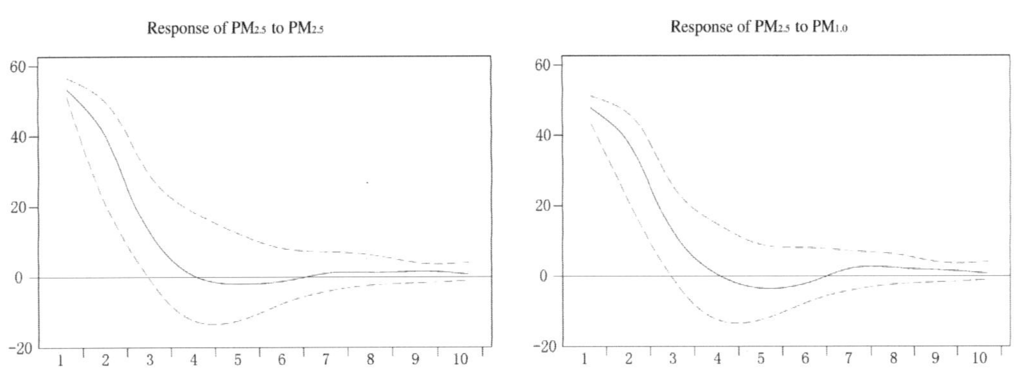

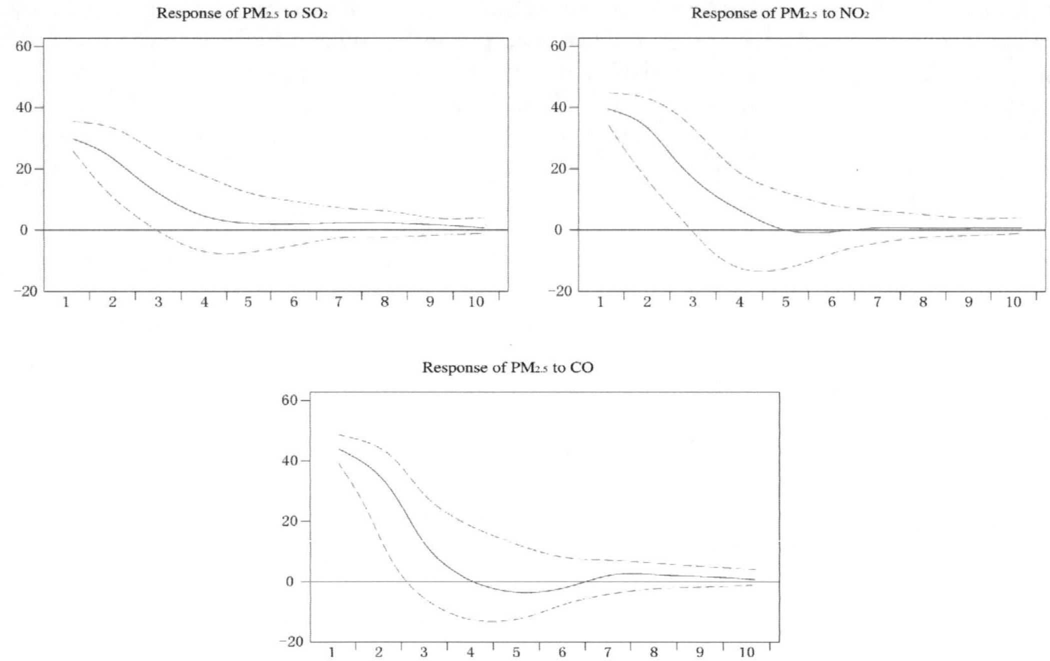

Figure 1. Impulse Response Functions: Response of $PM_{2.5}$

Figure 1 shows the response of $PM_{2.5}$ when PM_{10}, SO_2, CO and NO_2 are given a change of a standard deviation. In the first period, a positive change of $PM_{2.5}$ gives itself the biggest shock, which reaches nearly 51.32%. With time goes on, the impulse decreases gradually and reaches the lowest (-2.88%) in the 5th period, then the shock begin to rise and in the 10th period reaches 0.42%. When PM_{10} changes a standard deviation, it has impact on $PM_{2.5}$ by 45% in the 1st period. Then the change of PM_{10} influence on $PM_{2.5}$ is falling to minimum -3.2% until the 5th period and reaches slowly 0.25% in the 10th period.

When SO_2 suddenly gets a shock of a positive standard deviation, the initial impact on $PM_{2.5}$ is 30.8%. From the 2nd period the impact is reduced until reaches to a minimum of 1.98% in the tenth period. The impact of a standard deviation of NO_2 on $PM_{2.5}$ in the 1st period is 38.75%, after a rapid decline in the 4th period it reaches 3.33%. In next few periods the impact of changes in NO_2 is negative, but in the 8th period it returns to the positive and remains at the level of around 0.4%. A positive impact of CO on $PM_{2.5}$ in the 1st phase is 43.03% and has a rapid decline to 7.8% in the 3rd period, in the later three periods the impact is negative. In the 7th period it returns to 0.44% and have change little in the later periods, remaining 0.63% in the 10th period.

It is evident that in the 1st period $PM_{2.5}$'s reactions to other variables are all relatively large, the biggest impact is that by itself, which reaches 51.32%. As time goes by, the influence of variables on $PM_{2.5}$ is reduced, but the degree of reduction is not the same. Generally speaking, in the first four periods the impact of the changes in each variable on $PM_{2.5}$ is large and decreases quickly over time. Starting from the 5th period, other variable's impact on $PM_{2.5}$ is relatively small, which marginally increases until the 10th time period. This can be understood utilizing a superimposed effect: Today's $PM_{2.5}$ is not only affected by today's SO_2, CO and NO_2, but by yesterday, the day before yesterday and even a few days ago. This is in accordance with the actual situation, when a certain substance concentration increase, it immediately exerts impact on $PM_{2.5}$, and over time this influence is gradually reduced but still plays a part.

From the IRFs results it is also obvious that among SO_2, CO and NO_2, CO imposes the biggest impact on $PM_{2.5}$ in the 1st period. The increase of CO in the air will lead to $PM_{2.5}$'s immediate and sharp increase, and the impact of NO_2 takes the second place, followed by the influence of SO_2 on

$PM_{2.5}$ which is relatively small. Their impact on $PM_{2.5}$ however is declining over time, and the impact of CO decreases faster. In the tenth period, the influence of SO_2 on $PM_{2.5}$ is still larger (1.98%), the impact of CO on $PM_{2.5}$ is 0.63%, and NO_2's influence is minimum (0.27%). Overall, CO has a greater impact on $PM_{2.5}$ in the initial stage, while the influence of SO_2 on $PM_{2.5}$ sustains a long time.

3.3. Variance Decomposition

Variance decomposition is used to study the contribution of other variables to the variance variation of $PM_{2.5}$. Results of Variance decomposition are as follows:

Table 1. Variance decomposition results of $PM_{2.5}$ (%)

Period	S.E	$PM_{2.5}$	PM_{10}	SO_2	CO	NO_2
1	51.32	15.75 (1.53)	12.03 (1.47)	36.08 (3.57)	34.81 (2.99)	1.32 (0.80)
2	63.55	17.77 (2.36)	11.51 (2.25)	34.00 (4.54)	30.85 (3.33)	5.85 (2.03)
3	66.56	18.60 (2.87)	10.54 (2.43)	33.53 (4.99)	28.11 (3.66)	9.20 (3.00)
4	68.01	18.35 (3.09)	10.23 (2.44)	32.58 (5.22)	28.62 (4.29)	10.20 (3.499)
5	68.68	18.08 (3.17)	10.11 (2.47)	32.10 (5.47)	29.55 (4.70)	10.14 (3.63)
6	68.95	17.96 (3.21)	10.04 (2.49)	32.01 (5.67)	29.91 (4.82)	10.07 (3.65)
7	69.09	17.90 (3.23)	10.00 (2.51)	32.14 (6.05)	29.95 (4.91)	10.06 (3.61)
8	69.19	17.86 (3.25)	9.97 (2.54)	32.14 (6.05)	29.95 (4.91)	10.06 (3.61)
9	69.25	17.83 (3.27)	9.95 (2.55)	32.19 (6.22)	29.94 (4.95)	10.06 (3.60)
10	69.30	17.81 (3.29)	9.94 (2.57)	32.23 (6.38)	29.94 (5.00)	10.05 (3.58)

As shown in Table 1, the variance change of $PM_{2.5}$ can be separately decomposed to these factors consisting of $PM_{2.5}$, PM_{10}, SO_2, CO and NO_2, giving us a detailed analysis of these factors contribution to $PM_{2.5}$'s variance changes: In the 1st stage, $PM_{2.5}$ by itself explain the change of variance accounting for 15.75%, with PM_{10} accounting for 12%, SO_2 explaining 36.08%, CO explaining 34.8% and NO_2 with 1.3%. As time goes on, $PM_{2.5}$ variance change by itself shows slow decrease, and in the 10th period stabilizes at 17.8%; With time passes by, $PM_{2.5}$'s variance change explained by PM_{10} is in slow decline and falls to 9.9% in the 10th period; The change of $PM_{2.5}$'s variance caused by SO_2 has always been at the highest level and reaches 32.2% in the 10th period; $PM_{2.5}$'s variance change caused by CO has also been gradually decreased but still maintains the higher level (30%); Meanwhile, the variance change of $PM_{2.5}$ explained by NO_2 has gradually increased from the original 1.3% and in the 10th period been steadied at around 10%.

Just as IRFs, SO_2 and CO explain the most of the variance of $PM_{2.5}$. NO_2's impact on the variance change of $PM_{2.5}$ gradually increases over time, this may be because the proportion of SO_2 and CO on $PM_{2.5}$ change in the variance is reduced, which leads to an increase the proportion of NO_2's explanation. On the whole, the analysis results of variance decomposition are in accordance with that in the impulse response functions, SO_2 and CO have the greater influence on $PM_{2.5}$, followed by the effects of NO_2.

4. Conclusions

In this paper, Vector auto-regression (VAR) mode together with Granger causality tests, Impulse Response Functions and variance decomposition is utilized to verify the interaction between $PM_{2.5}$, PM_{10}, SO_2, CO and NO_2. Results are in the following: Granger causality tests manifest that there exists causal relationship between $PM_{2.5}$, PM_{10}, SO_2, CO and NO_2. SO_2, CO and NO_2 are influential factors of $PM_{2.5}$ and PM_{10}, namely, the increase of $PM_{2.5}$ and PM_{10} concentration in Beijing has certain relationship with the increase of concentration of SO_2, CO and NO_2. Impulse response results show that the response of $PM_{2.5}$ to a shock in CO is positive and large in the short term, while the elasticity of $PM_{2.5}$ with respect to SO_2 increases over time. As time progresses, $PM_{2.5}$ becomes more sensitive to the shift in SO_2. Meanwhile, Variance decomposition results indicate that $PM_{2.5}$ is more closely related to CO in short term, while SO_2' influence accounts for a higher proportion in the long run. Besides, NO_2's contribution to $PM_{2.5}$ is an added process, the effect of NO_2 began to be embodied as the extension of lag. These findings provide new view to analyze the dynamic influence factors of $PM_{2.5}$ and contribute to a better understanding of haze and its relationship with sustainable development.

Acknowledgements
This research was supported in part by the National Natural Science Foundation of China under Grant No.71271018 & No. 71531001.

References

[1] Hu D and Jiang J 2013 A study of smog issues and pm2.5 pollutant control strategies in china *J. Environ. Prot.* **04** 746-752
[2] Contini D, Cesari D, Donateo A, Chirizzi D and Belosi F 2014 Characterization of pm10 and pm2.5 and their metals content in different typologies of sites in south-eastern italy *Atmosphere* **5** 435-453
[3] Cox L A 2013 Caveats for causal interpretations of linear regression coefficients for fine particulate (pm2.5) air pollution health effects *Risk Anal.* **33** 2111–2125
[4] Chow J C, Bachmann J D, Wierman S S, Mathai C V, Malm W C and White W H 2002 Visibility science and regulation *J. Air Waste Manage. Assoc.* **52** 628-713
[5] Cheng L, Sandhu H S, Angle R P and Myrick R H 1998 Characteristics of inhalable particulate matter in alberta cities *Atmos. Environ.* **32** 3835-3844
[6] Sawant A A, Na K, Zhu X and Iii D RC 2004 Chemical characterization of outdoor $pm_{2.5}$ and gas-phase compounds in mira loma California *Atmos. Environ.* **38** 5517-5528
[7] Landis M S, Norris G A, Williams R W and Weinstein J P 2001 Personal exposures to $pm_{2.5}$ mass and trace elements in baltimor md usa *Atmos. Environ.* **35** 6511-6524
[8] Othman J, Sahani M, Mahmud M, and Ahmad M K 2014 Transboundary smoke haze pollution in malaysia: inpatient health impacts and economic valuation. *Environ. Pollut.* **189** 194-201
[9] Zheng S, Cao C X and Singh R P 2014 Comparison of ground based indices (api and aqi) with satellite based aerosol products *Sci. Total Environ.* **488** 398-412
[10] Adomavicius G, Bockstedt J and Gupta A 2010 Modeling supply-side dynamics of it components products and infrastructure an empirical analysis using vector autoregression *Inform. Syst. Res.* **23** 397-417
[11] Love I and Zicchino L 2006 Financial development and dynamic investment behavior: evidence from panel var *Q. Rev. Econ. Financ.* **46** 190-210

Enhanced Thermal Performance of Mosques in Qatar

A Al Touma[1] and D Ouahrani[1,*]

[1]Department of Architecture and Urban Planning, College of Engineering, Qatar University, P.O. Box 2713, Doha, Qatar

*Corresponding Author: djamel@qu.edu.qa

Abstract. Qatar has an abundance of mosques that significantly contribute to the increasing energy consumption in the country. Little attention has been given to providing mitigation methods that limit the energy demands of mosques without violating the worshippers' thermal comfort. Most of these researches dealt with enhancing the mosque envelope through the addition of insulation layers. Since most mosque walls in Qatar are mostly already insulated, this study proposes the installation of shading on the mosque roof that is anticipated to yield similar energy savings in comparison with insulated roofs. An actual mosque in Qatar, which is a combination of six different spaces consisting of men and women's prayer rooms, ablutions and toilets, was simulated and yielded a total annual energy demand of 619.55 kWh/m^2. The mosque, whose walls are already insulated, yielded 9.1% energy savings when an insulation layer was added to its roof whereas it produced 6.2% energy savings when a shading layer was added above this roof. As the reconstruction of the roof envelope is practically unrealistic in existing mosques, the addition of shading to the roof was found to produce comparable energy savings. Lastly, it was found that new mosques with thin-roof insulation and shading tend to be more energy-efficient than those with thick-roof insulation.

1. Introduction

Thorough analysis of the buildings' thermal performances is crucial in mitigating the increasing energy consumption. Among different types of buildings, religious spaces require special function and operation treatments as the indoor environment necessitates thermal and visual comforts, tranquillity and peace.

Qatar, as most states of the Arabian Gulf region, suffers from a hot climate due to the long-lasting summer season where the outdoor temperature reaches a daily mean of 37.7°C [1]-[2]. Air-conditioning (AC) is typically deployed from April through October, leading to tremendous electrical energy demands needed to control indoor spaces [3]. In fact, residential and non-residential buildings in Qatar consume over 60% of the total energy produced in the country [4]. According to the Ministry of Awqaf and Islamic Affairs [5], there are 1,227 mosques in Qatar. Since mosques compose a significant share of these buildings, any reduction in their thermal energy demands will result in significant energy savings that could be used in other more productive sectors.

In religious spaces, AC systems are usually kept on most of the time, which leads to limited indoor environment at considerable energy costs [6]. Since reduction changes in the AC systems are presumed to at least affect the indoor environment and the occupants' thermal comfort, reducing the mosque energy demands by enhancing its thermal envelope is considered as a better alternative solution.

The reduction of the mosque thermal energy demand using enhanced building envelope has been the scope of just a few researchers. For instance, the annual energy consumption of several mosques in the Kingdom of Saudi Arabia (KSA) was measured and an annual average of 182 kWh/m^2 was found [7]. Similarly, Al-Hamoud et al. [8] conducted experimental measurements on the total energy consumed in three different mosques in KSA. They found that air-conditioning cooling needs comprise up to 79% of the total energy of the mosque, where the rest is attributed to lighting and ventilation demands. In addition, it was found that a mosque with roof and wall thermal insulation layers added to the envelope consumed 93.5 kWh/m^2/year, much lower than another mosque with no insulation that consumed 201.5 kWh/m^2/year. Although their measurements were conducted on incomparable buildings, the results reflect the importance of envelope enhancement in yielding energy savings. Moreover, according to Budaiwi et al. [7], more than 20% of the annual cooling energy in mosques could be saved upon the addition of thermal insulation to its envelope.

Generally, mosques' walls in Qatar are thermally insulated as they comprise a considerable portion of the envelope surface area, whereas the roofs are left uninsulated. This study introduces a novel architectural shading design of mosques' roofs that is anticipated to yield similar energy savings in comparison with the conventional addition of insulation layers. In fact, since redesigning the roof construction layers is unpractical and consequently extremely expensive, the proposed model is applicable to both new and existing buildings.

2. Methodology

2.1. Description of the Mosque

This study considers an actual mosque in Qatar. The mosque design details, such as its footprint and construction material, were provided by the Ministry of Awqaf and Islamic Affairs [5]. The mosque, whose footprint is shown in figure 1, is composed of six different spaces.

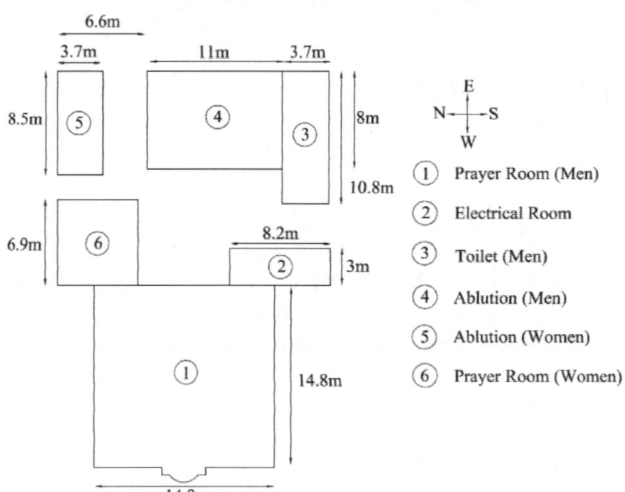

Figure 1. Schematic showing the mosque different rooms/spaces

As seen in figure 1, the selected mosque mimics an actual design of a prayer building in Qatar; it consists of a large prayer room for men, a smaller prayer room for women, separate ablution and toilets for men and women as well as an electrical and storage room. The spaces have different heights ranging from 3.8m to 5.75m, as well as different window and door sizes. The entire building has three entrances in the north, east and south orientations. The mosque design takes into consideration the necessities of any house of prayers in Islam. For this reason, women's privacy is ensured by their own entrance in the north direction that is blocked from the courtyard. In addition, their ablution space has no windows. On the other hand, the mosque considers the men's necessity to attend prayer times, and hence a larger prayer room of 218.3m^2 floor area was built for them. The door to the men's prayer room is on the wall facing east for worshippers coming from the other two entrances towards the courtyard. Both men and women prayer rooms consist of spaces with high ceiling, reaching 5.75m.

Furthermore, the mosque also considers the Mihrab, which is the niche that indicates the direction of the Kaaba in Makkah. The Minbar, which is the place where the leader of the prayer addresses the worshippers, has been built in the men's prayer room on the wall facing the west direction. Note that one meter high parapets have been built on the top of all six spaces, as seen in figure 2. These devices are used to prevent rain from reaching the vertical facades; it is accumulated in one corner and drawn down into sewage systems. Although the parapets are not tall, they are drawn on the spaces so that their minor effects on the buildings thermal performance is accounted for in the simulation. Note that this simplified mosque does not consider several architectural designs that might have little to no effect on its thermal performance.

The mosque envelope construction materials, starting from the outer layers, are as follows:

- Walls: 5mm gypsum plaster – 15mm sand/cement finish – 150mm thick hollow block – 100mm thick solid block – 50mm thick polyurethane insulation – 150mm thick hollow block – 15mm sand/cement finish (U-value = 0.57 W/m^2·K).
- Roofs: 15mm sand/cement finish - 300mm lightweight concrete (U-value = 1.85 W/m^2·K).
- Floors: 600mm lightweight concrete (U-value = 0.95 W/m^2·K).
- Windows: 6mm clear pane – 13mm airgap – 6mm clear pane (U-value = 1.95 W/m^2·K).
- Doors: 50mm thick solid wood door (U-value = 2.4 W/m^2·K)

2.2. Simulation Model

To be able to simulate the mosque and study its thermal performance, energy simulation software was deemed necessary. OpenStudio 1.13.0 software was selected for this purpose; it uses the EnergyPlus engine for the calculation of the thermal loads using the heat balance method, has a graphical user interface, and may use drawing software such as SketchUp Pro 16, the one picked in this study to build the mosque. A screenshot of the entire mosque as drawn on SketchUp software for simulation purposes is shown in figure 2.

After drawing the mosque, the construction envelope of the wall, roof and floor was designed layer by layer and attributed to the corresponding surfaces. Each space was assigned a unique zone of the same building for distinct air-conditioning control. The internal load consisted solely of electric lighting and occupancy as no electrical equipment seemed necessary. According to the Mosque Development Committee [9], the lighting power density in mosques should be in accordance with those outlined in ASHRAE 90.1-2007 Standards [10]. Hence, the lighting power density in the men and women's prayer rooms was set to 2.4 W/ft^2 (25.8 W/m^2) corresponding to religious buildings with worship pulpit, while it was set to 0.69 W/ft^2 (7.4 W/m^2) at the toilets and ablutions consistent with typical restrooms. However, these lighting power densities are not continuous; lights are rather assumed to be turned on and off depending on the occupancy schedule. Regarding the mosque occupancy, its density was set to 0.8 person/m^2. This value was found reasonable in this study, as it falls within the occupancy range of a typical mosque (0.62-1.00 person/m^2) [11]. The worshippers are considered to have a metabolic rate of standing people with light to medium activity, hence dissipating 160 W (2.0 met), according to ASHRAE Standard 55-2010 [12]. As for the occupancy schedule, Muslim traditions indicate five prayers per day, which may differ from one day to another. However, the variations in prayer times are in minutes and can be simplified into five fixed prayers. Note that this simplification will have no implication on the thermal analysis of the building. The appointed prayer times that were selected are 5AM, 12PM, 3PM, 5PM and 7PM. Note that each prayer time was estimated to last for 45 minutes, which is more than the time needed for typical prayers considering the arrival and departure periods of the worshippers. The occupancy schedule has been allocated to lighting demands as lights were turned off outside the prayer time intervals. Similarly, the occupancy schedule was attributed to the infiltration loads due to open doors as worshippers are coming into or out of the spaces. The infiltration rate was set to 1 ACH during occupied times and 0.5 ACH during unoccupied times [13].

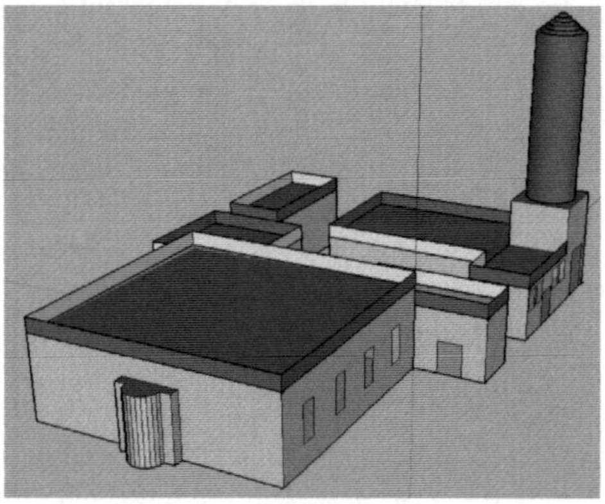

Figure 2. The entire mosque as drawn on SketchUp.

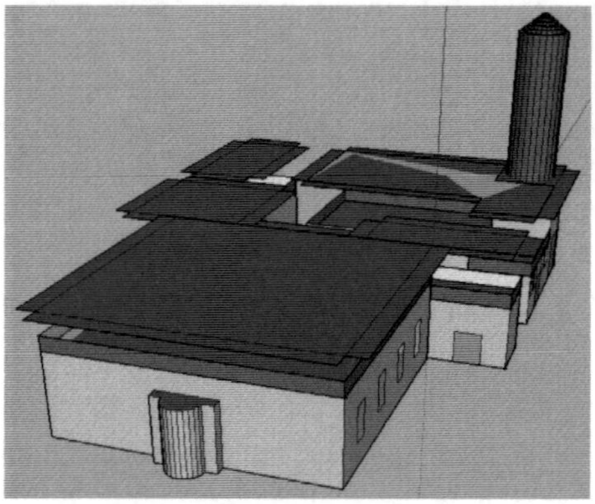

Figure 3. Mosque with shading installed on the roof.

Due to the hot weather conditions in Qatar almost throughout the year, the windows were assumed to be fully-closed. The *Turn On Ideal Air Loads* option was used in order to estimate the site total heating and cooling thermal energy demands without having to define any detailed air-conditioning system [14]. Also, the cooling set-point temperature was set to 24°C all day throughout the year since literature has proven little tendency of turning off the cooling systems in mosques even outside the occupancy intervals. Yet, one of the limitations of this software is that it assumes dual set-point schedule, and hence a heating set-point temperature is required [14]. In other words, a heating schedule should be inputted even into the zones that are expected to require cooling only. Research has shown that this problem is mitigated by selecting a heating schedule that initiates at a practically unachievable low temperature, such as 12°C.

2.3. Mitigation Methods for Enhanced Thermal Performance

The first alternative that is considered to mitigate the high total energy demand of the mosque and reduce its cooling loads is the addition of thermal insulation to the roofs. Similar to the ones installed on the walls, a 50mm and 100mm thick polyurethane insulation layer is installed on the roofs between the concrete and the sand/cement finish. In comparison with the old 1.85 $W/m^2.K$ U-value, the new roofs with 50mm and 100mm insulation layers have U-values of 0.56 $W/m^2.K$ and 0.33 $W/m^2.K$, respectively.

The second alternative considered in this study is the installation of shading device above the roofs of all spaces of the mosque. The purpose of this mitigation method is to come up with a better solution to save energy in comparison with the addition of insulation layer, and is applicable to existing mosques and cheaper in terms of initial capital cost. The new shading device is added at a 1.5m above the parapet level and covers the entire spaces roof surface area. The distance above the parapet height serves as a space for the heated air to be convected away through the passing air currents, hence removing heat accumulated within this area. In addition, these shadings are extended one meter in all directions for installation purposes. A screenshot of the updated mosque as drawn on SketchUp is shown in figure 3. The Casting Real-World Shadows feature on SketchUp was used to check whether the distance of the roof shading is able to shade most of the roof surface area, and the visualization seemed satisfactory. Both updated simulations were run and further analyses were conducted to compare their corresponding energy savings.

2.4. Outdoor Conditions

The hourly weather data for the State of Qatar, which was recorded during 2015 in Doha, was used in this simulation [15]. Since this study deals with the enhancement of the roof envelope, the weather parameters that mostly affect its thermal performance are the outdoor dry-bulb temperature and the Global Horizontal Irradiance (GHI). According to the weather file, the average temperature per day ranges between 35.7°C in August and 18.3°C in January. On the other hand, the total GHI per day ranges between 7632.6 Wh/m^2 in May and 3258.6 Wh/m^2 in December. This data seems to match with recent results of the work done by Perez-Astudillo and Bachour [1]-[2].

3. Results and Discussion

Upon running the simulations, results were extracted for detailed analysis.

3.1. Site Total Annual Energy Demand

The results that were of interest are the site total annual energy demands composed of the zones' air system cooling energies, lighting energies, infiltration heat gain energies and peoples' total heating energies. Table 1 shows the total energy demands for the base-case mosque, mosque with 50mm roof insulation, mosque with 100mm roof insulation and mosque with roof shading.

Table 1. Site total energy demand for the aforementioned simulation scenarios.

Simulation	Base-case Mosque	Mosque with 50mm Roof Insulation	Mosque with 100mm Roof Insulation	Mosque with Roof Shading
Site Total Energy (kWh/m^2)	619.55	563.43	550.72	581.07
Energy Savings (%)	-	9.1	11.1	6.2

As seen in table 1, the base-case mosque with neither roof insulation nor roof shading has a total site energy demand of 619.55 kWh/m^2. Al-Hamoud et al. [11] found that typical mosque energy demand ranges between 186.2 kWh/m^2 and 201.5 kWh/m^2, based on actual data recordings. However, their mosques consisted of one prayer room only. Considering the fact that this study inspects the energy loads of six thermal zones consisting of two men and women's prayer rooms along with their ablutions and toilets, the total energy demand was found to be reasonably larger.

Upon the installation of the 50mm insulation layer on the spaces roofs, the mosque total energy demand decreased to 563.43 kWh/m^2, thus yielding 9.1% energy savings. Higher energy savings of up to 11.1% were found when the insulation layer thickness was increased to 100mm. On the other hand, by solely installing the shading above the spaces roofs (refer to figure 3) and without any insulation layers, 6.2% energy savings were found with a total energy demand of 581.07 kWh/m^2. This highlights the importance of installing shading on the mosque roof, as it turns out to save comparable energy in comparison with the addition of insulation layers.

3.2. Monthly Total Cooling Demand

The mosque monthly total cooling demands for all four simulated cases are shown in figure 4.

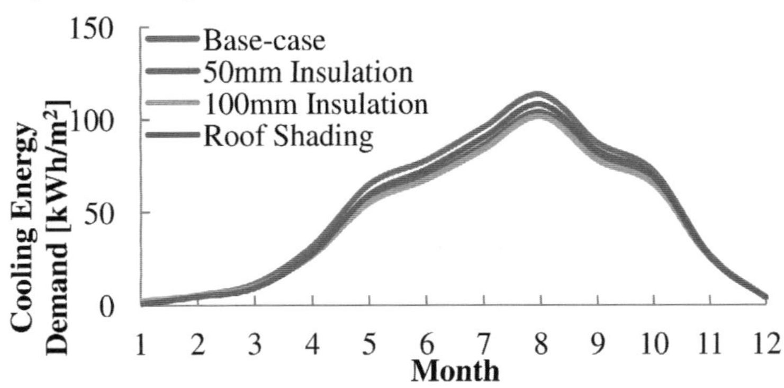

Figure 4. Mosque total monthly cooling demand.

As shown in figure 4, the total cooling energy demand was at maximum during the summer season between July and September and minimum during the winter season between December and February. For the base-case mosque with no roof insulation or shading, the mosque cooling demand ranged between 2.19 kWh/m^2 in January and 113.59 kWh/m^2 in August. For the mosque with 50mm and 100mm roof insulation, the cooling demand followed a comparable trend but with peak values of 103.75 kWh/m^2 and 101.44 kWh/m^2, respectively. The mosque with roof shading had a less intense reduction in the peak load, which reached 108.07 kWh/m^2, along with attenuation of the monthly demands for the rest of the year.

The reasons behind the results of table 1 and figure 4 could be explained as follows. The addition of polyurethane insulation layer that has a very low conductivity (0.02-0.04 W/m.K) leads to significant rejection and resistance to thermal heat passage through the roofs. However, the walls of all spaces are already insulated, which limits the effect of insulation addition on the roofs and leads to moderate energy savings compared to other studies [7]-[8]. In addition, all mosque spaces tend to have larger floor-to-ceiling heights in comparison with other office or residential spaces. For instance, the space heights of this mosque range between 3.80m and 5.75m, whereas it is only 2.8m in typical office spaces [16]. This fact also decreases the effect of enhancing the roof envelope due to its low ratio of surface area with respect to the entire envelope surface area. On the other hand, the addition of the shading reflects all direct solar radiation before it reaches the roof surface. Yet, it still allows diffuse and reflected radiation rays to hit different surfaces. In addition, it contains a 1.5m distance from the roof, which allows air currents to remove any heat accumulated within this region. Nonetheless, it is not made of insulative material, which means that it slows down heat passage due to its thermal mass more than it rejects it off its surface. For this reason, the installation of the shading had a moderate effect on the thermal envelope and led to 6.7% and 6.2% energy savings in the site cooling and total energy demands, respectively. Hence, these results reflect the ability to save energy easily in existing mosques with the simple installation of roof shading.

3.3. Men's Prayer Room

Going deeper into thermal zones, the men's prayer room was picked as it represents the major contributor towards the mosque total site energy demand. Hence, the percentage breakdown of energy demands for the base-case simulation for the men's prayer rooms is calculated. As anticipated, more than half (54%) of the space energy demands are attributed to cooling. Since Qatar is a country with hot weather conditions throughout the year, cooling is anticipated to have a substantial share of the total load. The rest of the total load is internal and divided between people, lighting and infiltration. As mentioned before, worshippers visit the mosques five times per day throughout the year. With an occupancy density of 0.8 person/m^2, the internal load caused by occupants reached around 30% of the total loads for men's prayer room directly followed by infiltration load (9%) that is subject to the same occupancy schedule. Lastly, lighting internal load had the lowest contribution, accounting for only 7%. To study the thermal demands of the men's prayer room only, their total annual energy end uses were extracted and presented in table 2.

Table 2. Total annual energy end uses of the men's prayer rooms for the studied cases.

Simulation	Cooling (kWh/m^2)	Lighting (kWh/m^2)	Occupants (kWh/m^2)	Infiltration (kWh/m^2)	Total (kWh/m^2)
Base-case Mosque	288.17	41.77	158.18	44.91	533.03
Mosque with 50mm Roof Insulation	239.14	41.77	158.18	42.14	481.23
Mosque with 100mm Roof Insulation	227.42	41.77	158.18	41.33	468.70
Mosque with Roof Shading	243.58	41.77	158.18	42.66	486.19

As seen in table 2, the men's prayer room total annual energy demand is 533.03 kWh/m^2, 288.17 kWh/m^2 of which are attributed to cooling needs whereas the rest is internal loads. The addition of the 50mm and 100mm insulation layers decreased the annual energy demands to 481.23 kWh/m^2 and 468.70 kWh/m^2 respectively due to the decrease in the cooling and infiltration loads. For the case when shading is added on the top of the mosque, the men's total energy demand decreased to 486.19 kWh/m^2. Since the lighting and occupants' internal loads are inputs to the simulated model, they are not affected by any changes in the envelope and consequently have a fixed annual thermal load of 41.77 kWh/m^2 and 158.18 kWh/m^2, respectively.

As shown in table 2, it could be noticed that the performance of the mosque different spaces, and more specifically the men's prayer rooms, is fairly similar when the roof is insulated with 50mm polyurethane and when it is shaded. These results re-highlight the benefits of installing the shadings on the thermal energy demands of different spaces.

4. Roof Insulation and Shading of New Mosques

In this section, the mosque has been reconstructed with enhanced roof envelope through the addition of 50mm roof insulation as well as roof shading. The model was re-simulated and results were as follows.

The site total energy demand turned out to be 539.62 kWh/m^2, yielding 12.9% energy savings in comparison with the base-case mosque with neither roof insulation nor roof shading. The results seem interesting as this mosque produced more energy savings in comparison with a mosque with 100mm roof insulation (refer to table 1). As for the mosque total cooling demand, the mosque had a more intense reduction where the peak decreased from 113.59 kWh/m^2 (refer to figure 4) to 99.87 kWh/m^2 along with reduced loads for the rest of the year. Hence, 13.5% of the annual cooling demand was saved. Similarly, the results for the men's prayer rooms were extracted for analysis.

Table 3. Total annual energy end uses of the men's prayer rooms with insulated and shaded roofs.

	Cooling (kWh/m^2)	Lighting (kWh/m^2)	Occupants (kWh/m^2)	Infiltration (kWh/m^2)	Total (kWh/m^2)
Men's Prayer Room	216.39	41.77	158.18	40.97	457.32

As seen in table 3, the annual cooling energy demands of the men's prayer room decreased significantly from 288.17 kWh/m^2 to 216.39 kWh/m^2 when the roofs construction was insulated and shaded. Similarly, the infiltration loads decreased as the envelope was more tightened. This construction envelope led to 14.2% energy savings in the total annual energy demands of the men's prayer room. Consequently, large amounts of energy could be saved if such conservation strategies are adopted in new mosques.

5. Conclusion

Mosques are plentiful in the Arabian Gulf, and their increasing consumption of energy is something that contradicts with the countries' future energy-conservation visions. In Qatar, most mosques' walls are already insulated, thus limiting the capability of building envelope enhancement in reducing energy demands. In this study, a novel design consisting of roof shading is compared to the conventional addition of insulation layer to the roof, and results were compared to a base-case mosque with neither roof insulation nor roof shading.

An actual mosque consisting of six spaces built according to the design and building envelope characteristics provided by the Ministry of Awqaf and Islamic Affairs in Qatar was simulated on Openstudio and estimated to demand 619.55 kWh/m^2/year. As the space walls are already insulated, the addition of thin and thick insulation layers to the mosque roofs was found to yield 9.1% and 11.1% energy savings. On the other hand, the addition of shading to the mosque roof singlehandedly produced 6.2% energy savings. As the roof reconstruction is practically unrealistic, these results highlight the importance of roof shading as a mitigation measure for saving comparable energy in existing mosques. As for new mosques, it was found that roofs built with 50mm insulation and shading outperforms those built with 100mm insulation and tend to produce 12.9% energy savings.

This study provides practical mitigation methods against the increasing energy consumption in existing and new mosques as the addition of the insulation layers in the roof is cumbersome and expensive. Future work includes further studies on the optimization of the parameters related to the installation of the shading device on the roof of the mosque combined with life-cycle cost analysis.

6. Acknowledgment

The authors would like to acknowledge Qatar University for their support in providing the necessary tools for the completion of this study.

References

[1] Perez-Astudillo D, Bachour D. Variability of measured global horizontal irradiation throughout Qatar. *Solar Energy*. 2015; **119**:169-78.
[2] Bachour D, Perez-Astudillo D. Ground measurements of global horizontal irradiation in Doha, Qatar. *Renewable energy*. 2014; **71**:32-6.
[3] Al-ajmi FF, Hanby VI. Simulation of energy consumption for Kuwaiti domestic buildings. *Energy and buildings*. 2008; **40(6):**1101-9.
[4] Ayoub N, Musharavati F, Pokharel S, Gabbar HA. Energy consumption and conservation practices in Qatar—A case study of a hotel building. *Energy and Buildings*. 2014; **84**:55-69.
[5] Ministry of Awqaf and Islamic Affairs. Online Resource: http://www.islam.gov.qa/. Accessed on 20 Oct 2016.
[6] Tham KW. Conserving energy without sacrificing thermal comfort. *Building and Environment*. 1993; **28(3)**:287-99.
[7] Budaiwi IM, Abdou AA, Al-Homoud MS. Envelope retrofit and air-conditioning operational strategies for reduced energy consumption in mosques in hot climates. In *Building Simulation* 2013 (pp. 1-18). Tsinghua University Press, co-published with Springer-Verlag GmbH.
[8] Al-Homoud MS, Abdou AA, Budaiwi IM. Mosque energy performance, part II: monitoring of energy end use in a hot-humid climate. *Engineering Sciences*. 2005; **16(1)**.
[9] Mosque development committee, Abu Dhabi mosque development regulations, Volume 2, Design.
[10] Standard AS. Energy standard for buildings except low-rise residential buildings. ASHRAE/IESNA Standard. 1999;90(1).
[11] Al-Homoud MS, Abdou AA, Budaiwi IM. Assessment of monitored energy use and thermal comfort conditions in mosques in hot-humid climates. *Energy and Buildings*. 2009; **41(6)**:607-14.
[12] Standard AS. Standard 55-2010:"Thermal Environmental Conditions for Human Occupancy"; ASHRAE. Atlanta USA. 2010.
[13] Budaiwi IM. Envelope thermal design for energy savings in mosques in hot-humid climate. *Journal of Building Performance Simulation*. 2011; **4(1)**:49-61.
[14] OpenStudio 1.9.0 Basic Workflow Guide. Online Resource: https://www.openstudio.net/ Accessed on 23, Oct 2016.
[15] White box technologies. Online Resource: http://weather.whiteboxtechnologies.com/. Accessed on 25 Oct 2016.
[16] Al Touma A, Ghali K, Ghaddar N, Ismail N. Solar chimney integrated with passive evaporative cooler applied on glazing surfaces. *Energy*. 2016; **115**:169-79.

An Input and Output Analysis of the Quaternity-Dominating Energy Engineering Model from China's Countryside

Xing Long Xie[1,*] and Wei Xian Xue[2]

[1]Economics & Management Institute, Xi'an University of Technology, Xi'an, China
[2]Scientific Research Office, Xi'an University of Technology, Xi'an, China

*Corresponding e-mail:long86886@aliyun.com

Abstract. The aim of this study is to qualitatively and quantitatively explore an energy engineering model termed quaternity-dominating pattern emerging in North China's countryside. This study finds methane produced in this model serves household activities such as cooking, inducing reduction of coal or biomass spending, which otherwise would provoke air pollution, water loss and land erosion, and ultimately leading to ecological environment betterment. Additionally, this project generates byproducts, biogas liquids and residuals, which can, as a category of fertilizer, can promote straightening of fertility preservation capacity and improvement in the chemical and physical quality of land as well as increasing crop output and quality. This study also finds this engineering could encourage social stability via efficiently allocating bucolic surplus labor during winter and successful running this engineering project would trigger an increase of scientific and technological qualifications for rural citizens. Moreover, cost-profit analysis indicates this pattern can allow one rural home to obtain access to a hygienic energy resource of biogas in the yearly volume of 375m^3, generate annual net earnings of US$3458.82 and make investment return in about 2.73 years. Especially for poverty-stricken areas, this energy engineering project enjoys high values and great significance, which can lift these impoverished areas from poverty both in economics and energy. The paper concludes with pointing out practical proposals on launching and operating this energy engineering project.

1. Introduction

Energy has evolved into a major element required to reduce poverty, increase the standard of living and promote social-economic advancement [1].The steadfastly swift pace of economy gives rise to strong energy demand and spending having connections with challenges and negative consequences among developing nations. A considerable number of them have suffered a severe energy plague for decades, with some counties lacking of access to electricity. More seriously, rural and remote areas in those middle-low income countries have undergone painful crises including inadequate or no public energy provision [2]. Typically, among vast rural regions of bulky population in China, a representation of high economic growth among the developing world, energy is consumed at a substantial volume, with the power composition taking on diversity. Massive direct combustion of biomass such as stalks and fuel woods is central to power spending in the rustic life [3]. Alarmingly, this extensive pattern has provoked an enormous devastation to the ecological environment as well as low levels in energy efficiency [4]. This, coupled with economic poverty in areas such as loess hilly maintains among Gansu, Ningxia and Shaanxi, has posed formidable threats to sustainable development. Thus, it is getting increasingly imperative to seek out an energy engineering system, which should comply with principles from the sustainable development ideology and must pave the way to upgrading of power resources, alleviating poverty, and aiding economic growth and environmental conservation.

Over the past decades, China has developed a variety of engineering modes, among which the quaternity-dominating pattern (QDP) denoting four in one, boasts a typical sustainable energy development and moves at a fast pace as it has reversed energy shortage and ecological exacerbation disturbing an expanse of rural areas in north China. Notwithstanding achieving expected results, this energy engineering model remains considerable problems. Thus, it is particularly imperative to make an overall analysis of this mode to formulate countermeasures in favor of extending in other rural areas to lift their households from energy poverty. Thus, this study represents a high worth for northwest China's poor loess areas, which, with tattered landforms and dense gullies, has degenerated into the most severe area of water loss and soil erosion [5], undergoing a cruel fragility in the ecological environment, and, where bearing lack of energy, families here have to draw on local resources such as stalk, wood and weed for cooking and heating purposes [6], so having provoked a growing ecological deterioration [7].

The aim of this paper is to afford evidence on further extension and application of QDP via exploring this model from the approach of input and output and illuminating its comprehensive benefits including economic, social and ecological. Our analysis uses the newest data, which gives our study to capture current markets. We consider five main aspects: (1) construction costs, (2) production expenditures, (3) economic earnings, (4) ecological benefits, and (5) technological and knowledge improvement. These aspects allow us to address many of energy engineering questions set out in existing energy efficiency studies. Our analysis thus provides a broad evaluation of profit-cost in the energy area and contributes to the growing body of literature on rural energy by answering key questions.

Our major findings are as follows: (1) The quaternity-dominating pattern, which incorporates a methane-generating pit, a pig or poultry house, a sunshine greenhouse and a household court to shape into an ecological system of energy under the closed condition, generating methane, planting crops such as cucumber and cultivating livestock such as pigs and ultimately enhancing an upgrading of energy spending, an alleviating of poverty and an accomplishing of ecological balance among the loess hilly areas, which have undergone energy famine, economic poverty and environmental deterioration. (2) Cost-profit analysis indicates that the suggested QDP can allow one rural home to obtain access to a hygienic energy resource of biogas in the yearly volume of 375m^3, generate annual net earnings of US$3458.82 and make investment return in about 2.73 years. (3) By adopting this energy engineering, poor rural populations are freed from the fear which is based on want. Therefore, the energy engineering scheme takes on a promising prospect and merits priority development in energy poverty areas such as central Gansu's loess mountains.

The remaining of this paper is organized as follows. Section 2 contains an outline concerning QDP. Cost estimation is presented in section 3. Section 4 covers benefits and effects arising from this energy engineering model. Conclusions and policy proposals are in section 5.

2. Profile of QDP

This structure (see figure 1), built in a rural household court, arrays a piggery, a toilet and a greenhouse in the form of word "one", with a methane tank constructed below the piggery, a feed inlet connected to the entrances of human & animal droppings and the discharge hole stretching into the indoor. The methane-generating pit is responsible for fitting together planting, livestock breeding, agricultural production and rural life. Here, human and domestic animal droppings, after fermenting, yield gases that can produce CO_2 via combustion as to push an increase of output in fruit and vegetables. Being superior fertilizers, such dregs and liquids can lead to a reduction of both chemical fertilizer and agricultural agent adoption. With regard to the solar greenhouse, it can make agreeable illumination, temperature, and humidity for the methane tank and the swine pen or fowlery. For the toilet, its role is to furnish the biogas-generating pit with excrement, a category of raw material for the purpose of fermentation. In addition to providing raw materials, the swine pen is devoted to enhancing its temperature by increasing the CO_2 density in the interior of the green house.

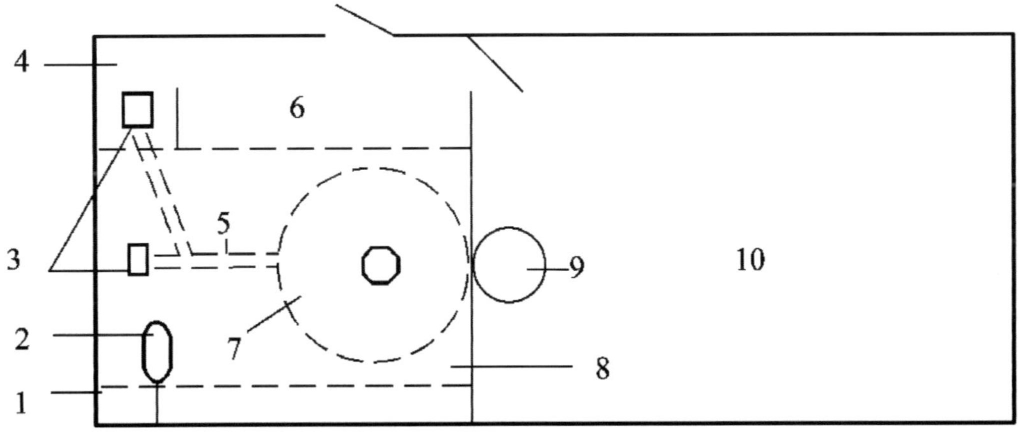

Figure 1. System layout of QDP.

Note: 1. fence, 2. trough for spilling water, 3. orifice for charging stock, 4. toilet, 5. inlet pipe for charging stock, 6. passage, 7. methane tank, 8.swine pen, 9. discharging hole, and 10.solar greenhouse.

3. Cost estimation

3.1. Cost for greenhouse construction

It is suggested that the steel frame greenhouse should be adopted because it agrees with anti-seasonal crops such as vegetables during north China's cold winter due to its high quality of heat preservation. This hothouse covers 350 m^2, spanning a length of 35 meters from east to west and internal width of 10 meters. Almost all, except walls, are made into a full steel framework, covered with drip-proof, anti- aging PO films and cotton wadding. Thus, it is strong and durable. For its expense, it is expected to reach US$ 3804 (26248.7RMB) by engineering contract, calculated by virtue of data online, whose proprietor, Agricultural Firm of Wise Monarch located in Shandong province, has been dedicated to engineering construction for greenhouses since 1990 and now enjoys high prestige nationally. Concerning its lifespan, it is estimated to be 20 years [8].

3.2. Cost for biomass-generating pit building

Regarding creating a methane-generating pit, its volume has become a primary concern since a relatively undersized pit may cause a weak output of gas, failing to meet purposes of production and life, and, a comparatively large one could contribute to a lower ratio in the methane-making. Hence, in conformity with the criterion of 1.5-2 M3 per capita [9], a tank with the area ranging from 8 to 10 m^3 can satisfy à five member family. Besides, considering winter weather, it is suggested creating a methane-generating pit in the volume of 10 m^3. Such a pit, according to the prevailing market of China, is predicted to cost US$577.9 and is computed at 15 years of lifespan.

3.3. Cost for pig house establishment

Above the methane pit and bordering the greenhouse on one end in the east or west, a pigpen with an area between 12 and 14 m^2, accommodating 8 swine or so, should be built. Furthermore, it ought to take the form of plastics canopy, which can take advantage of sunlight to cause the internal temperature to rise, thus benefiting pig growth, even over a winter period. Such a 12 square meter pigpen is estimated at about US$ 521.74, on the footing of unit price US$43.4782 [10].

3.4. Pig-feeding expense

Direct expenditure, which encompasses pigs, feed-stuff, medical treatment and so forth, usually making up 70% of total expenses, is crucial to the cost budget of pig–breeding, and is subject to ever-increasing prices of production factors such as piglets and fodder. It is found that the direct cost for a pig comes in at around US$214, grounded on the current market and data stemming from current literature [11]. Assuming raising 16 swine, annual investment reports US$3424.

3.5. Crop-planting outlay

For vegetable cultivation inside a greenhouse, the cost is made up of three parts: material and services, labor, and land. Concerning an engineering project of QDP, the two latter expenses could be written off because family members as employees are engaged in related activities such as agricultural production and land is of free use, dictated by Chinese legislation. Consequently, the cost for crop planting inside the greenhouse has reduced to the material and service outlay, consisting of two items:(1) the direct production expense, mainly arising from purchase of seed, farming film, pesticide, fertilizer, and maintenance services; and (2) the indirect expense, mainly involving fixed asset depreciation and sales. Owing to difficulty of direct production data availability, Shandong province's number as an agent value was used for cucumber cultivation. It is at US$ 0.7736/m^2 in 2012.

Considering production material inflation 1.625%, the current direct cost for cucumber cultivation is figured to be US$0.8251/m^2. For a 320m^2 planting area, the direct production cost totals to US$ 528 for cucumber, based on two plantings per year.

3.6. Outlay for complement facilities

Complement facilities, including that straw mulch, covering a 350m^2 area, with an online offer quoted at US$1.087/m^2 and 1.5 years of lifespan, represents an annual outlay of US$253.63; and a rolling machine, priced at US$217.4 online and 10 years in life [12], is computed at an annual charge of about US$21.74, cost around US$275.37 annually.

4. Benefit Assessment

4.1. Economic effect

Annual revenue: the temperature inside the greenhouse's canopy, ranging from 10 to 30 centigrade fits in with two plantings per year. To take cucumbers for example, a planting area of 320m^2 inside a 350m^2 canopy can lead to 7116.8 kg in output, on the unit productivity at 11.12kg/m^2 [13]. According to the online offer of USD0.58/kg , total earnings can hit about US$ 4127.7(see table 1).

For breeding, one year can witness the annual aggregate output of 16 pigs in two lots, with each pig around Kg115 [14], valued US$ at the unit price US$2.3188/16.0RMB/kg online, pig raising can make US$ 4266.6. A 10 m^3 methane tank can produce 375m^3 of gas annually, equivalent to 267.75 kg in standardized coal or 669.4kg of ordinary coal (considering heat efficiency), valued at US$ 83.4 calculated on the unit price of 124.64US$/ton.

So, the planting, breeding and biogas system leads to an annual income up to US$8477.7 or so, which, reduced by US$5018.88 (annual cost), yields a net profit of about US$3458.82(see table 1), implying it would take approximately 2.73 years to make investment return for investors. Provided that some "renowned, specific, superior and fresh" animal breeds or immune-to-season vegetables or fruit are introduced, this model would perform more beneficially. Hence, each household would be thoroughly lifted from poverty and lead a better life by means of this engineering project.

Table 1. Investment Cost and Earnings. (Unit: US$)

Item	Greenhouse	Methane tank	Pig house	Complement facilities		Production		Sum
				Mulch	Rolling machine	planting	Breeding	
Life span	20	15	15	1.5	10	-	-	-

One-time cost	3804	577.9	521.74	380.45	217.4	528	3424	9453.4
Annual average cost	190.2	38.53	34.78	253.63	21.74	1056	3424	5018.88
	263.51			275.37		4480		
Annual income	83.4(methane value)			-		4127.7 4266.6		8477.7
Net annual income	-			-		-		3458.82

4.2. Ecological benefit

The biogas manure has been recognized as a category of high quality organic fertilizer, created through ferment of substantial quantities of microbes. It contains (1) nutritional substances supplying nutritional elements such as nitrogen, phosphorous and potassium for crops;(2) trace elements in the form of ions; (3)some hormones such as somatotropin, gibberellins and abscisic; (4) aminoacid and vitamin B [15]. Accordingly, the manure of biogas can lead to plenty of positive outcomes following adequate fertilization. One significant consequence is to increase the vegetable output, improve its nutritional quality and enhance its commercial value. Take mini-cucumbers for example, superior to chemical and farmyard manures, biogas liquids and residues are found to push the cucumbers' vitamin C content up to 48.63mg/100g, dissolvable sugar up to 22.116% and nitrate at 90.74 mg/kg [16]. Another positive effect is reflected by biogas liquids soaking seeds, which would bring about a rise of 5%-10% in the germination percentage and 10%-15% in the seedling rate. Third, methane fertilizer can result in rising of organic matter contents, the permeable rate and the water retention and fertility preservation capacity and, ultimately, improvement in the chemical and physical quality of land [17]. Finally, biogas both liquids and residuals, including amino acid and antibiotic factor, are endowed with functions such as getting rid of insects and bacteria, including aphids, which might be destroyed through biogas liquids being sprayed on leaves, thus contributing to reduction of plant diseases and insect pests, further to less utilization of chemical pesticides and eventually to production of healthier vegetables.

Moreover, methane serves, as a category of fuel, rural household activities such as cooking and heating, consequently resulting in a reduction of coal or biomass spending and ecological environment betterment. Generally, coal burning could provoke massive pollutants, including SO_2, which exerts harmful effects on respiratory tissues through causing serious illness like pneumonia and edema and on the environment through generating acid rain, NO, which is a damaging substance of ozone as well as physical body problems such as bronchitis and lung cancer, and Hg0, which turns out to be deleterious. Additionally, firewood combustion would induce forest and plant damage, water loss and land erosion, and environmental deterioration. All those negative consequences could be alleviated by establishment of the proposed engineering project.

4.3. Social benefit

The QDP energy engineering could promote social stability via efficiently allocating surplus labor and thus transforming the bucolic idle status in winter when there is no farming work for rural inhabitants to undertake, a worrying scene concealing possible social risks, just as voiced by the saying "idle hands are the origin of evils". Next, excrement can flow into the methane-manufacturing pit at whatever time and this would thoroughly reverse the unhygienic, stinking condition of a rural court plagued by flies and mosquitoes and cut off channels of harmful bacterium spreading to create a healthy, living environment productive to local citizens. Third, successful running of this energy engineering project for achieving expected results, lays great requirements on rural citizens, who must, by training and doing, get familiar with and put into practical application various agricultural knowledge and skills ranging from related equipment operations, to biogas generating, to greenhouse vegetable and pig cultivating, to the planning of investment and capital budget. This, undoubtedly, would trigger an increase of scientific and technological qualifications for rural citizens, an establishment of the correct idea that sustainable development in agriculture is carried out through advanced technologies, and a mastering of modern agricultural marketing skills. Finally, just as addressed above, due to methane liquids and residuals(healthy and nutritional) substituting for chemical manures to fertilize soils and plants or being utilized to feed pigs, the model is able to enhance the supply of healthy foodstuffs such as greener vegetables and healthier pig meat as well as increasing farmer incomes.

5. Conclusion and proposal

5.1. Conclusion

The paper is devoted to exploring QDP, a novel energy engineering project, which, delivering ecological, economic and social benefits, is capable of liberating rural areas from double poverty in energy and economy under environmental conservation to achieve sustainability of energy, ecological, economic and social development. Here are new insights. First, the QDP of energy engineering is proposed for sample areas, which conforms to the ideology of sustainable development. This fresh energy engineering requires that each family should install and put into use a green house with an area of 350 m^2, a methane tank of 8-10 m^3, a warm awning of 10 m^2 for livestock and birds, a toilet of 1 m^2 and an energy-efficient kang stove. Such a project, with a one-time investment at about \$9453.4, is projected to yield an annual net income of around US\$3458.82 and make investment return in roughly 2.73 years. Next, the fresh model can generate marsh gases as a major form of energy to be adopted by rural households, contributing to a reduction in the burning of coal and firewood and ultimately relieving environmental contamination. Third, livestock and human excrement entry into the methane tank at whatever time allows the farm households to be freed from bad smells and contagion, thereby facilitating physical health improvement. Furthermore, biogas liquids and residuals, employed as a category of fertilizer, can promote straightening of fertility preservation capacity and improvement in the chemical and physical quality of land as well as increasing crop output and quality. More significantly, this engineering model can improve overall qualifications of rural citizens by imbuing modern agricultural knowledge, marketing skills of produce and the advanced ideology into their minds.

5.2. Proposal

To initiate and spread an engineering scheme concerning the quaternity-dominating mode, the primary assignment ought to develop the interest and awareness of fresh energy consumption of local residents, trapped in the mire of following the beaten way by the force of habit. Primarily, a certain number of innovation-minded villagers, who are competent and pioneering, should be organized to make an on-the-spot investigation of successful sites in practicing this new energy mode to stimulate their enterprise-seeking desire. Then, those natives need to be imparted systematic knowledge associated with methane, planting, breeding, marketing and circular economy via inviting experts or technicians to teach them, enabling them to have corresponding techniques including management, which are a prerequisite for triumphant operations of this engineering. Third, it is particularly pressing to reinforce establishment of a service network, as the maintenance and use of each equipment in this project leads to great demands for technical services in each of these corresponding areas, and the operation of each activity such as fermentation, crop planting, livestock raising and marketing of products to be manufactured by this engineering puts new demands on professional service sectors for their growing complexity. Otherwise, such a project, however superior, would fail in achieving expected results, even exert a huge waste, for instance, it, whose building may cost considerable resources including funds and materials, couldn't generate any gain at all if certain equipment can't be operated or repaired when it goes wrong or a certain link such as fermenting has a problem, in the absence of expert services from technicians. Fourth, governmental grants such as low-interest loans and financial funds ought to be furnished for those rural households, not moneyed, since massive resources of manpower, material and finance are necessitated by constructing this engineering project. Eventually, remote rural areas such as loess hilly, suffering economic poverty and ecological fragility, the QDP energy engineering style, which can trigger a simultaneous fulfillment of energy, ecological, economic and social benefits and has been referred to as an event productive to lower-level individuals and households, should be legalized and implemented, as an item falling into the local government's rural development planning. Provided all those suggested proposals are pursued, this model will be increasing and bearing fruit among more rural regions to profit local families. It would exercise a far-reaching effect on rural energy composition upgrading. Significantly, this new-fashioned power engineering could merit extension in other developing nations such as Mongolia and North Korea.

Acknowledgment
The authors are grateful to financial support from the national social-scientific fund titled "study of harmonious development of production networks and ecological environment across the silk road belt" No.17BJL005.

References

[1] Ershad U K and Andrew R M 2016 *Renewable and Sustainable Energy Reviews* **62** 247-59.
[2] Barnes D F, Khandker S R and Samad H A 2011 **39** 894-904.
[3] Chen Y, Yang G and Sweenney S 2010 *Renew Sustain Energy Rev* **14** 545-9.
[4] Qiu H G 2015 *China Soft Science* **299** 28-38.
[5] Yang L Z, Wang G F, Wang A J, Yao Y H, Hu Y H, Li Y W, Liu Zh R, Guan H W and Liu J H 2016 *The Chinese Journal of Geological Hazard and Control* **227** 39-48.
[6] Li G Z and Dong D X 2010 *Renewable Energy Resources,* **28**,115-9.
[7] Li G Z, Nu S W and Yang Z 2008 *Journal of Natural Resources* **23** 15-24.
[8] Huang H 2016 *Gansu Agriculture* **18** 34-35 turning to 40.
[9] Sun B l 2008 *Chinese Journal of Eco-agriculture* **16** 1279-82.
[10] Zhang H P 2013 *Rural Animal Production Technology* **06** 21-3.
[11] Huang L Y 2015 *Guang Dong Agricultural Science* **16** 175-9.
[12] Liu X B 2012 *Agricultural Science & technology and Equipment* **08** 51-2.
[13] Yan F, Zou Zh, Dong J, Li J, Zhang Zh X, Wang Y B and Yang J J 2009 *Acta Agriculturae Boreali-occidentalis Sinica* **18** 272-5.
[14] Wang Z L 2012 Swine Industry Economic 01 7-9.
[15] Zhou Q, Wang L Y 2013 Northern Horticulture 17 192-5.
[16] Hao X J, Hong J P and Gao W J 2007 Soil and Fertilizers Sciences in Chin 05 40-3.
[17] Rivard C J, Rodriguez J B and Nagle N J 1995 Biochemistry and Biotechnology 52 125-35.

Novel Fluorinated Tensioactive Extractant Combined with Flotation for Decontamination of Extractant Residual during Solvent Extraction

Xue Wu[1], Zhidong Chang[1,*], Yao Liu[1] and Chol Ryong Choe[1, 2]

[1]Department of Chemistry and Chemical Engineering, School of Chemistry and Biological Engineering, University of Science and Technology Beijing, Beijing 100083, PR China
[2]School of application chemistry, Kim Chaek University of Technology, Pyongyang 999093, DPR of Korea

*Email: zdchang@ustb.edu.cn

Abstract. Solvent-extraction is widely used in chemical industry. Due to the amphiphilic character, a large amount of extractant remains in water phase, which causes not only loss of reagent, but also secondary contamination in water phase. Novel fluorinated extractants with ultra-low solubility in water were regarded as effective choice to reduce extractant loss in aqueous phase. However, trace amount of extractant still remained in water. Based on the high tensioactive aptitude of fluorinated solvent, flotation was applied to separate fluorinated extractant remaining in raffinate. According to the data of surface tension measurement, the surface tension of solution was obviously decreased with the addition of fluorinated extractant tris(2,2,3,3,4,4,5,5-octafluoropentyl) phosphate (FTAP). After flotation, the FTAP dissolved in water can be removed as much as 70%, which proved the feasibility of this key idea. The effects of operation time, gas velocity, pH and salinity of bulk solution on flotation performance were discussed. The optimum operating parameters were determined as gas velocity of 12ml/min, operating time of 15min, pH of 8.7, and NaCl volume concentration of 1.5%, respectively. Moreover, adsorption process of FTAP on bubble surface was simulated by ANSYS VOF model using SIMPLE algorithm. The dynamic mechanism of flotation was also theoretically investigated, which can be considered as supplement to the experimental results.

1. Introduction

Solvent extraction is extensively used in hydrometallurgy, waste-water treatment and biopharmacy, etc. However, extractant loss is an alarming problem after solvent extraction, which causes wastage of commercial solvent extractants as well as serious environmental contamination[1]. Many research suggested that the loss of organic extractants primarily depended on the solubility of the extractant in aqueous phase[2]. The novel fluorinated extractants with ultra-high hydrophobicity, such as tris(2,2,3,3,4,4,5,5-octafluoropentyl) phosphate, were proposed to be an effective method to reduce extractant loss in aqueous phase[3-4]. Nevertheless, trace amount of fluorinated extractants still remained in water. Hence, it is meaningful to seek for methods to further decontamination.

There are several industrial methods to remove organic solvents from water, such as physical heat treatment like distillation, chemical treatment like membrane adsorption and extraction, which are energy-intensive and contaminative. As a combination of chemical and physical method, floatation is on the basis of the fact that the contaminant in the wastewater itself are surface active and adsorbed on the gas-liquid two-phase interface by the bubbles floating in waste water. It has gradually become a new method of wastewater treatment because of low energy consumption and pollution. Compared with traditional physical and chemical treatments, floatation is particularly suitable for the low concentration substances and has been widely used as separation technology for the purification of oily wastewater[5-7]. Except for ultra-high hydrophobicity, high tensioactive is another prominent aptitude of fluorinated compounds[8]. Hence, flotation is assumed to be an appropriate method to separate and enrich the residual of fluorinated extractant.

In the present research, tris(2,2,3,3,4,4,5,5-octafluoropentyl) phosphate (FTAP) was chose as model fluorinated extractant. Novel fluorinated tensioactive extractant combined with flotation was proved to be a promising and eco-friendly strategy to tackle the problem of extractant residual during solvent extraction as shown in figure 1. Due to the complexity of the gas-liquid two-phase fluid and the interface phenomenon, floatation simulation was also conducted.

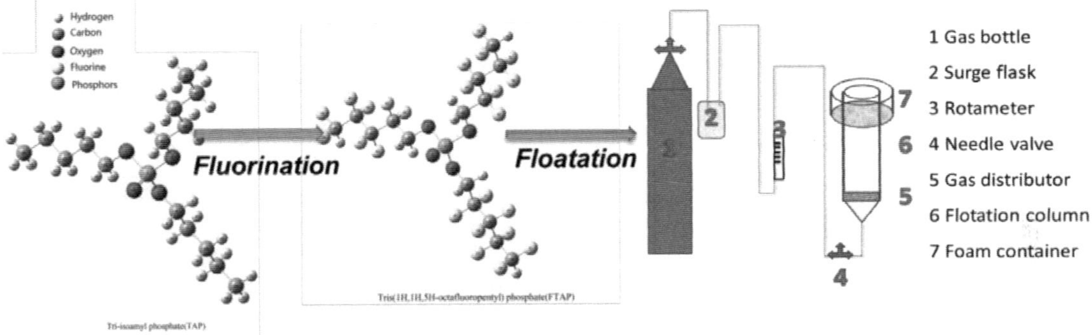

Figure 1. Schematic of research approach and flotation device.

2. Research methods

2.1. Measurement of surface tension

Surface tension of water solution containing different proportion of FTAP was determined by QBZY-1 automatic surface tensiometer.

2.2. Floatation process

FTAP dispersing in water phase was separated by floatation using CO_2. The amount of FTAP in solution was tested by fluorescence spectrophotometry (λ Ex=290 nm, Em=400nm). The separation efficiency E is defined as:

$$E\% = \frac{C_0 - C_t}{C_0} \times 100 \quad (1)$$

where C_0 stands for the initial FTAP concentration in the aqueous phase and C_t stands for FTAP concentration in the aqueous phase at certain time. The effects of operation time, gas velocity, pH and salinity of bulk solution on flotation performance were discussed.

2.3. Theoretical simulation

Adsorption process of FTAP on bubble surface was simulated by ANSYS VOF model using SIMPLE algorithm. The simulation was set in the 0.003 m × 0.005 m region. The quadrilateral grid length was 0.001 mm. The time step was 0.1 ns. The top boundary was set as the pressure outlet, and the other boundary was no slip wall. The finite volume method was used to discretize the partial differential equations, and the coupling relationship between pressure and velocity was solved by SIMPLE algorithm.

3. Results

3.1. Surface tension of FTAP solution at different volume ratio

The prerequisite of flotation is high tensioactive target materials. Therefore, the ability of FTAP to decline surface tension of water phase was firstly investigated. Surface tension curve of FTAP solution at different volume ratio was shown in figure 2. As shown in figure 2, it can be seen that with the increase of the content of FTAP, the surface tension of the FTAP solution decreases. When gas-liquid interface, i.e. bubbles, appears in the water, FTAP molecules dispersing in solution would tend to gather around bubbles, which benefits separation process.

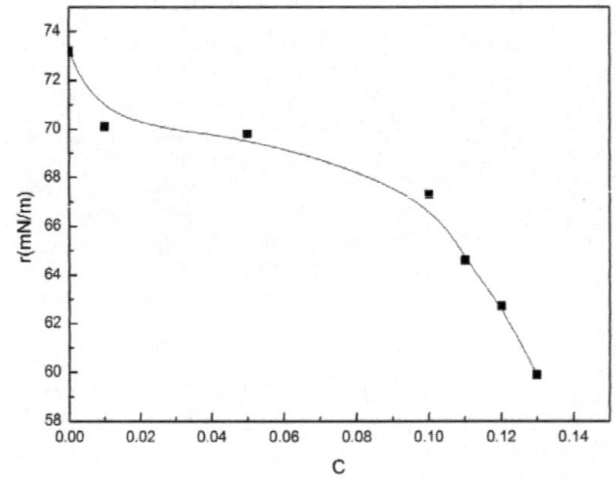

Figure 2. Surface tension curves of FTAP solution at different volume ratio.

3.2. Separation of FTAP in water by floatation

The residual of FTAP in water phase was separated by floatation using CO_2. The effects of operation time, gas velocity, pH and salinity of bulk solution on flotation performance were shown in figure 3, 4, 5, and 6, respectively. As shown in figure 3, with the increase of flotation time, the separation efficiency increases rapidly and gradually becomes flat after 10 mins, which implies that adsorption was close to equilibrium. From figure 4, it is obvious that separation efficiency is enhanced with increase of gas velocity, and maximizes at 12ml/min, indicating that increase of gas-liquid interface was beneficial for separation and enrichment of FTAP in water. Whereafter, separation efficiency decreases when gas velocity further increases. This could be the result that violent mix at the top of air flotation tower broke bubbles. Figure 5 shows the effect of pH on the separation efficiency at different gas velocities. By adjusting pH during experiments, the separation efficiency reaches the maximum when the pH is 8.7, indicating that slight alkalinity was conducive to maintaining stability of CO_2 bubble membrane. Figure 6 shows that the separation efficiency increases with the increase of NaCl concentration in a certain concentration range. When the NaCl concentration is about 1.5%, the separation efficiency reaches the maximum value. This indicates that the salinity of bulk solution had great effect on the formation of bubbles. Low concentrations of salt can prevent the rapid mixing of bubbles and increase the specific surface area of bubbles.

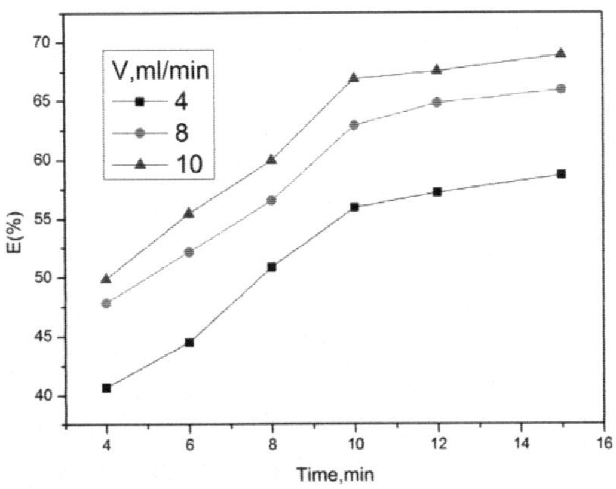

Figure 3. The effects of operation time on separation efficiency of FTAP.

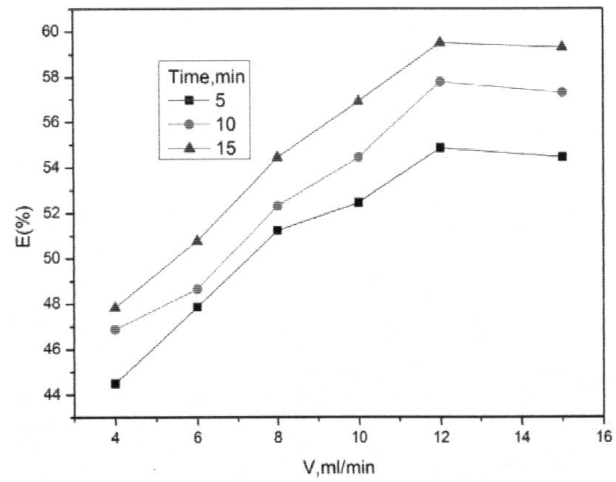

Figure 4. The effects of gas velocity on separation efficiency of FTAP.

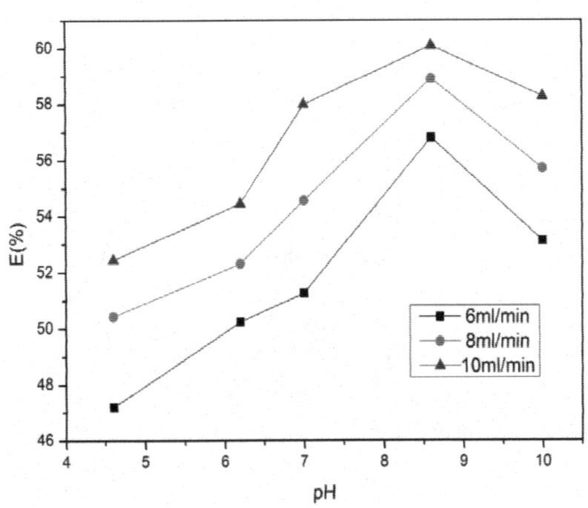

Figure 5. The effect of pH of bulk solution on separation efficiency of FTAP..

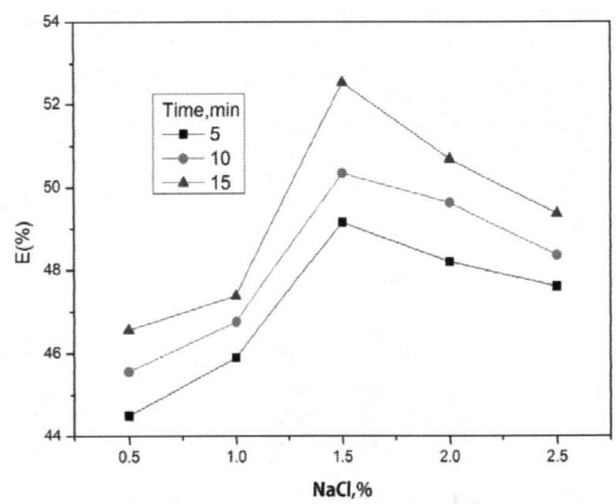

Figure 6. The effect of salinity of bulk solution on separation efficiency of FTAP..

3.2. Simulation of adsorption process during flotation

Based on ANSYS VOF method, flow volume change in flow field grid was chased to capture interface of two-phase flow[9-10]. The calculation method is small, accurate and easy to implement. The initial parameters of simulation of flotation bubble absorption were listed in table 1.

Table 1. Initial parameters of simulation of flotation bubble absorption

Parameter	Value
Initial diameter of bubble / mm	1
Density of CO_2 / (kg·m^{-3})	1.784
Density of FTAP / (kg·m^{-3})	1770
Density of water / (kg·m^{-3})	998.2
Viscosity of CO_2 / (Pa.s)	1.4932×10^{-5}

Viscosity of FTAP / (Pa.s)	5.89×10^{-4}
Viscosity of water / (Pa.s)	1.003×10^{-3}
Surface tension of water / CO_2 / ($N\cdot m^{-1}$)	0.0732[a]
Surface tension of FTAP / CO_2 / ($N\cdot m^{-1}$)	0.0181[a]
volume fraction of FTAP in system	1×10^{-4}

[a] Refer to the correlation coefficient of air

Figure 7 (a) shows the initial form of a fixed position bubble in model system. The volume fraction contours of FTAP and volume fraction curves of FTAP in centreline at 80 ns, 90 ns, 100 ns, 200 ns, 300 ns were presented in figure (b), (c), (d), (e), and (f), respectively. During the simulation process, FTAP molecules in water can spontaneously congregate around the bubble. The content of FTAP adsorbing on the surface of bubble increases with time and then decreases, which is in accordance with the experimental results. The bottom of the bubble adsorbs more FTAP than other position. It is because during the air flotation process, organic materials aggregate in the bottom of bubble on account of gravity. Hence, organic materials would slide from the bubble surface over a certain degree of time. The bubbles loading with FTAP should be scraped out opportunely in order to increase the separation effect. The simulation of the adsorption process can provide better understand to air flotation process and help to optimize the operating conditions.

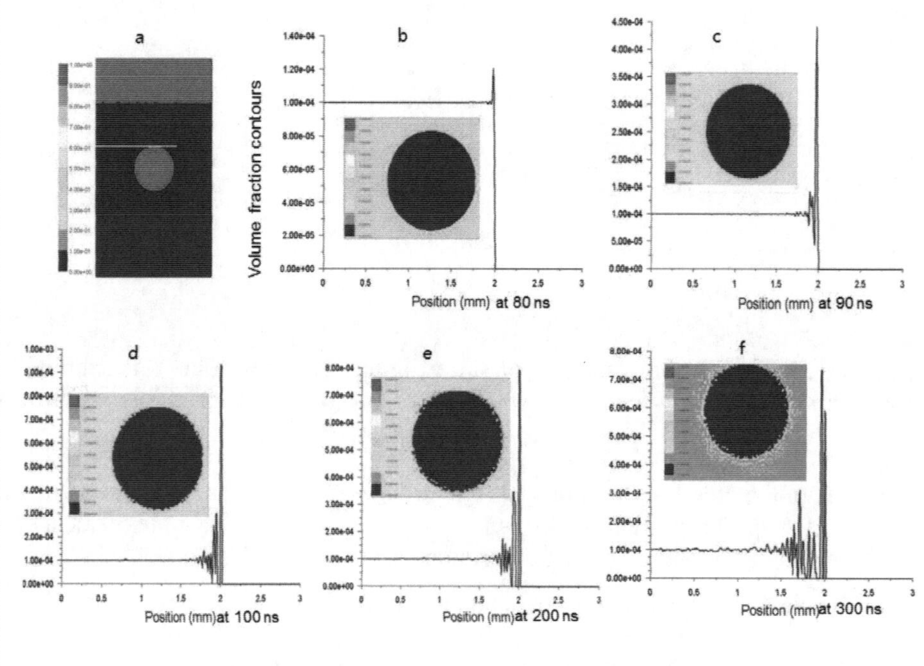

Figure 7. (a) Initial form of a fixed position bubble. The volume fraction contour of FTAP and volume fraction curve of FTAP in centreline at (b) 80 ns, (c) 90 ns, (d) 100 ns, (e)200 ns, (f)300 ns.

4. Conclusion

In summary, novel fluorinated extractant possess very low solubility in water and satisfactory extraction capability. Flotation was applied to further purify water phase on the basis of the high surface tension of fluorinated extractants. Operation time, gas velocity, pH and salinity of bulk solution were important factors affecting the separation efficiency. The residual of tris(2,2,3,3,4,4,5,5-octafluoropentyl) phosphate in water phase was separated by floatation using CO_2. The optimum operating parameters were determined as gas velocity of 12ml/min, operating time of 15min, pH of 8.7, and NaCl volume content of 1.5%, respectively. Simulated adsorption process of FTAP on bubble surface by ANSYS VOF model using SIMPLE algorithm was in accordance with the experimental results. Novel fluorinated tensioactive extractant combined with flotation was proved to be a promising strategy to tackle the problem of extractant residual during solvent extraction

Acknowledgments
This work was financially supported by the National Natural Science Foundation of China (No. 21276022).

References

[1] Liu C Cao J Shen W Ren Y Mu W and Ding X 2016 Liquid–liquid equilibria of quaternary system phosphoric acid/sulfuric acid/water/tri-n-butyl phosphate and the formation of extraction complex at 303.2 K *Fluid Phase Equilib* **408** 190-195.

[2] Semenov A Abramov A Sobol Y B Galieva Z N Galantsev A and Gerya V 2016 Preparation of Nd2O3 through the extraction separation of rare-earth concentrate recovered from phosphogypsum *Inorg. Mater* **52 (2)** 233-237

[3] Wu X Chang Z Blamo B J Yuan J Uslu H Li W and Ryong C C 2017 Strategy for extractant residual reduction: Experimental and computational investigation of fluorinated phosphate *Fluid Phase Equilib* **449** 167-174

[4] Luo YChang Z Wu X Uslu H Li W Hua C Liu S Sun C and Im S 2015 Efficiency of fluorinated alcohol for extraction of organic acid from its dilute aqueous solution: A model study *J. Fluor. Chem* **178** 260-265

[5] Sun X Chang Z Shen S Hu X and Liu H 2006 Effects of emulsion properties on recovering butyl acetate from wastewater of penicillin plant by solvent sublation *Colloids Surf. Physicochem. Eng. Aspects* **286 (1)** 8-16

[6] Rubio J Souza M and Smith R 2002 Overview of flotation as a wastewater treatment technique *Miner. Eng* **15 (3)** 139-155

[7] Zhu J Wu F Pan X Guo J and Wen D 2011 Removal of antimony from antimony mine flotation wastewater by electrocoagulation with aluminum electrodes *JEnvS* **23 (7)** 1066-1071

[8] Mohamed M and Mohamed A 2009 Synthesis and some properties of fluorinated cationic surfactants *Journal of surfactants and detergents* **12 (4)** 345-349

[9] Chakraborty D Guha M and Banerjee P 2009 CFD simulation on influence of superficial gas velocity, column size, sparger arrangement, and taper angle on hydrodynamics of the column flotation cell *Chem. Eng. Commun* **196 (9)** 1102-1116

[10] Cai X Chen J Liu M Ji Y and An S 2017 Numerical studies on dynamic characteristics of oil-water separation in loop flotation column using a population balance model *Sep. Purif. Technol.* **176** 134-144

Research on Application of Automatic Weather Station Based on Internet of Things

Chen Jianyun[1,*] Sun Yunfan[1] and Lin Chunyan[1]

[1]Fuzhou Meteorological Bureau, China

*385659847@qq.com

Abstract. In this paper, the Internet of Things is briefly introduced, and then its application in the weather station is studied. A method of data acquisition and transmission based on NB-iot communication mode is proposed, Introduction of Internet of things technology ,Sensor digital and independent power supply as the technical basis, In the construction of Automatic To realize the intelligent interconnection of the automatic weather station, and then to form an automatic weather station based on the Internet of things. A network structure of automatic weather station based on Internet of things technology is constructed to realize the independent operation of intelligent sensors and wireless data transmission. Research on networking data collection and dissemination of meteorological data, through the data platform for data analysis, the preliminary work of meteorological information publishing standards, networking of meteorological information receiving terminal provides the data interface, to the wisdom of the city, the wisdom of the purpose of the meteorological service.

1. A brief introduction about the Internet of Things

The Internet of Things (IoT) is the combination of communication network and Internet communication. There are two meanings: Firstly, the core and foundation of IoT is still the Internet, it's the extension and expansion of the network based on the Internet; Secondly, the user terminal extends to each item for information exchanging and communicating. Which means everything is related. Internet of Things has three key technologies in application, such as follows [1-2]:

a) Sensor technology. It's also the key technology in computer applications.
b) RFID tag, which is also a kind of sensor technology. RFID technology is a comprehensive technology which combines radio frequency technology and embedded system technology as a whole.
c) Embedded system technology. It is a complex technology combining computer hardware and software, sensor technology, integrated circuit technology and electronic application technology as one whole.

2. The IoT automatic station

2.1 Overview on the IoT automatic station

The IoT automatic station is automatic weather station based on the Internet of things, which fully embodies the real-time data interoperability, data fusion and other characteristics of the IoT technology. With the advent of the IoT, it will become increasingly widespread. The structure of the meteorological observation network is shown in Figure 1.

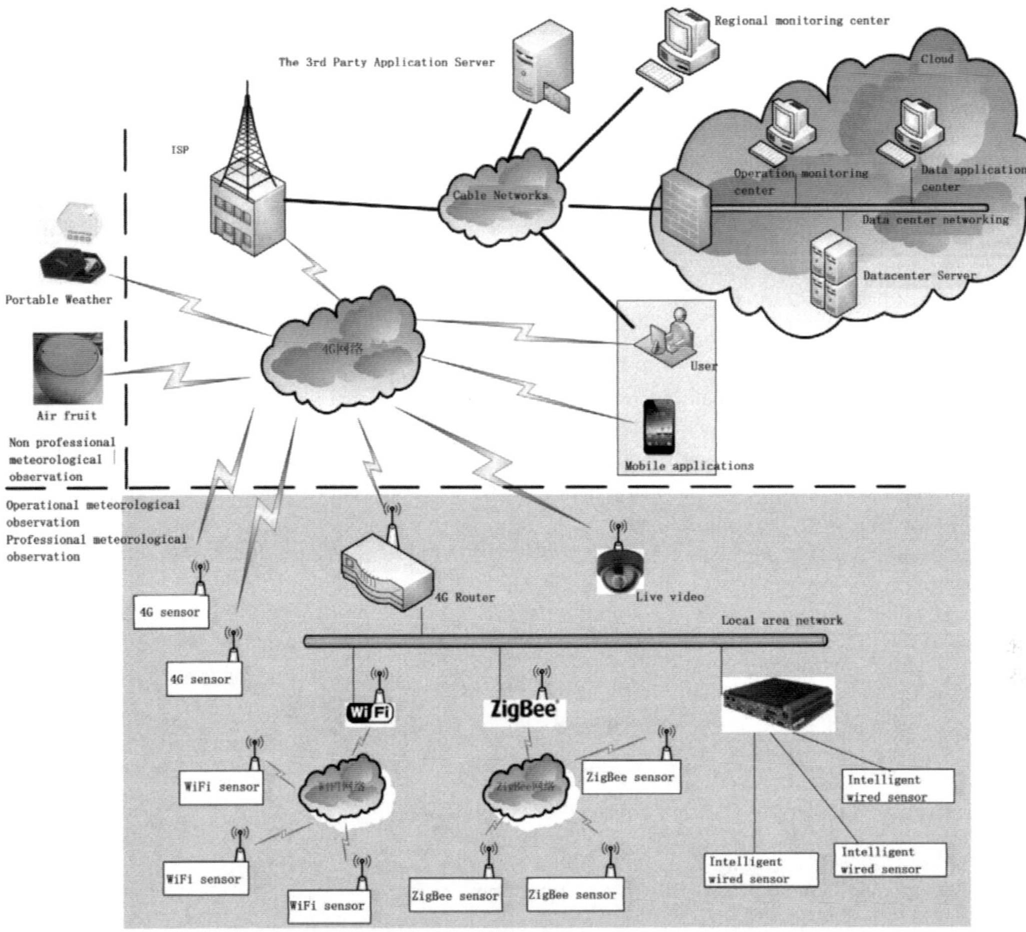

Figure 1. The structure of the meteorological observation network

2.2. Features of the IoT automatic station [3]

1) High quality meteorological data
   A. Intelligent perception: sensor self-calibration, sensor co-integration algorithm.
   B. Multi-quality control: single-factor quality control, multi-factor joint quality control, time and space consistency check.
2) Reliable meteorological equipment: intelligent sensor, wireless transmission, micro-power, self-organizing network.
3) Comprehensive monitoring information: self-test and self-diagnosis, multiple monitoring.
4) Efficient maintenance and protection: fault warning, alarm and diagnosis, operating status information.
5) Convenient deployment: equipment distributed, power supply diversification, transmission of wireless.
6) Standard interoperability: follow the Internet of things communication protocol standards and data formats, barrier-free dialogue.
7) Flexible Scalability: City, Province, Region, National.

3. Technical route of IoT automatic station

3.1. Technical overview

The architecture model that the existing automatic weather station uses is sensor plus collector mode.

The IoT automatic station should break it completely: eliminate power supply cables and communication cables, take the digital and independent power supply of sensors as foundation, and at the same time introduce Internet of things technology to compose the Internet-based automatic weather station.

3.2. Equipment composition

The IoT automatic station consists of power supply, wireless communication system, digital sensor, integrated controller and business software platform. Figure 2 shows the structure of the the IoT automatic station system [4].

1) Power supply system: use distributed power supply, digital sensors and integrated controller were equipped with independent solar power supply unit.
2) Wireless communication system: consists of two parts: WSN network and GPRS network, in which WSN network is based on Zigbee network.
3) The digital sensor is responsible for front-end data acquisition and preprocessing.
4) Integrated controller as a system gateway, responsible for the establishment of WSN network and summary of WSN data information, while taking the automatic station and business software platform data interaction.
5) The business software platform is responsible for the human-computer interaction of the IoT automatic station system.

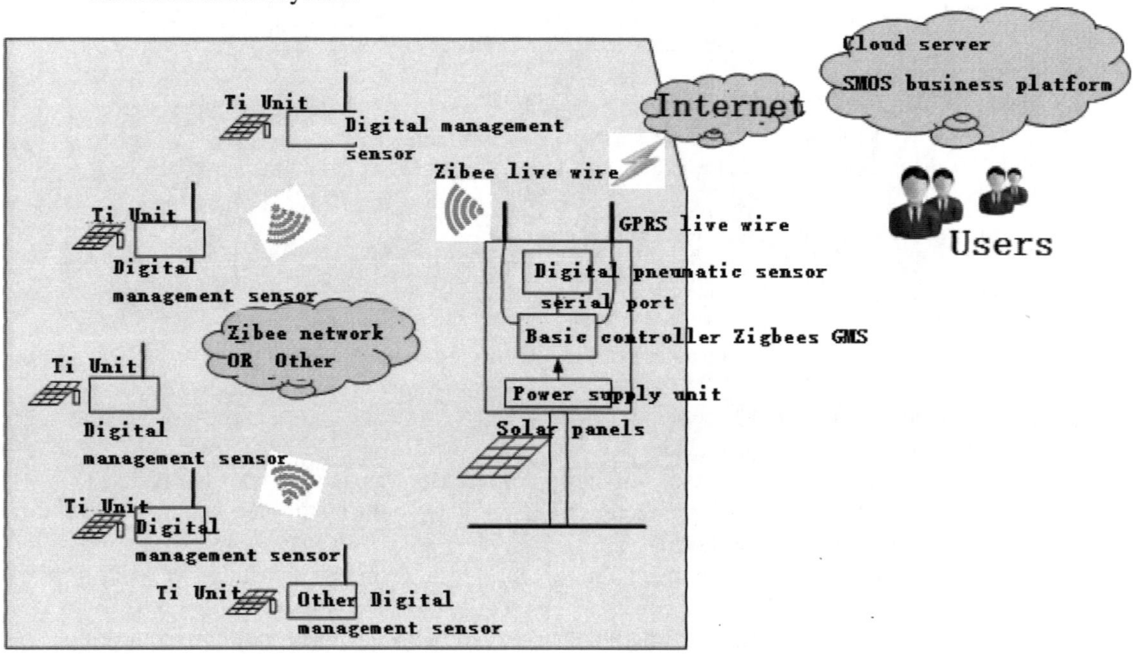

Figure 2. The structure of the the IoT automatic station system.

3.3. Device deployment

3.3.1 Cooperative deployment model of sensors

1) The front end takes the station as a unit, and a variety of intelligent sensors work together in the station. The location and deployment of the automatic weather station is compatible with the traditional station building mode, and the observation elements in the station are comprehensive and diversified.
2) The station uses the intelligent sensor + wireless sensor network architecture, intelligent sensor with a "dialogue" function and collaboration capabilities, increase the comprehensive quality control of the data fusion algorithm based on single sensor element quality control algorithm, to

improve the data quality; at the same time using data fusion technology for data calibration, improve data accuracy.

3) On the basis of network access mode of traditional automatic station, 4G/wifi and other network access methods are extended so as to make full use of the public resources of smart city network.

Figure 3.shows Cooperative deployment model of sensors.

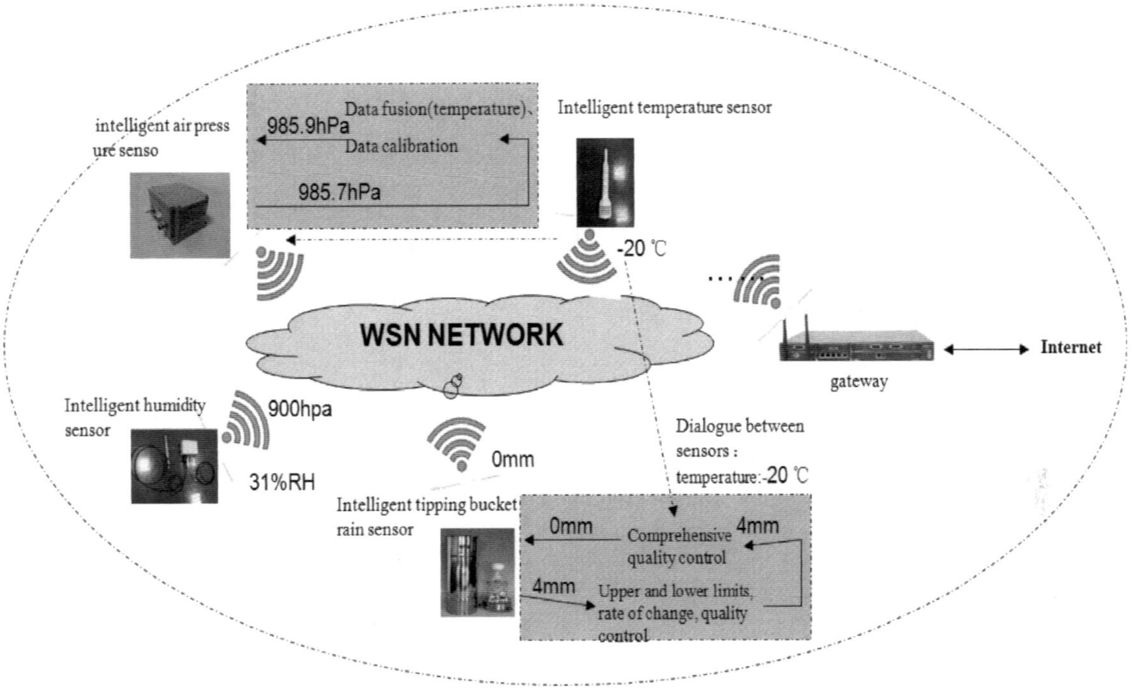

Figure 3. Cooperative deployment model of sensors.

3.3.2 Intelligent sensor independent deployment mode

1) The front end, with intelligent sensors as the unit, supports any element combination in space.
2) Single station deployment is simple, mobile fast and flexible, easy to fine, comprehensive monitoring [5].
3) Single station support 4G/wifi and other network access methods, smart city in the basic network coverage area, to achieve fast access.
4) Cloud server monitoring data information of different single stations can be classified and aggregated according to type, GPS location and application, relying on large data technology to process monitoring data and information fusion.

Figure 4.shows Intelligent sensor independent deployment mode.

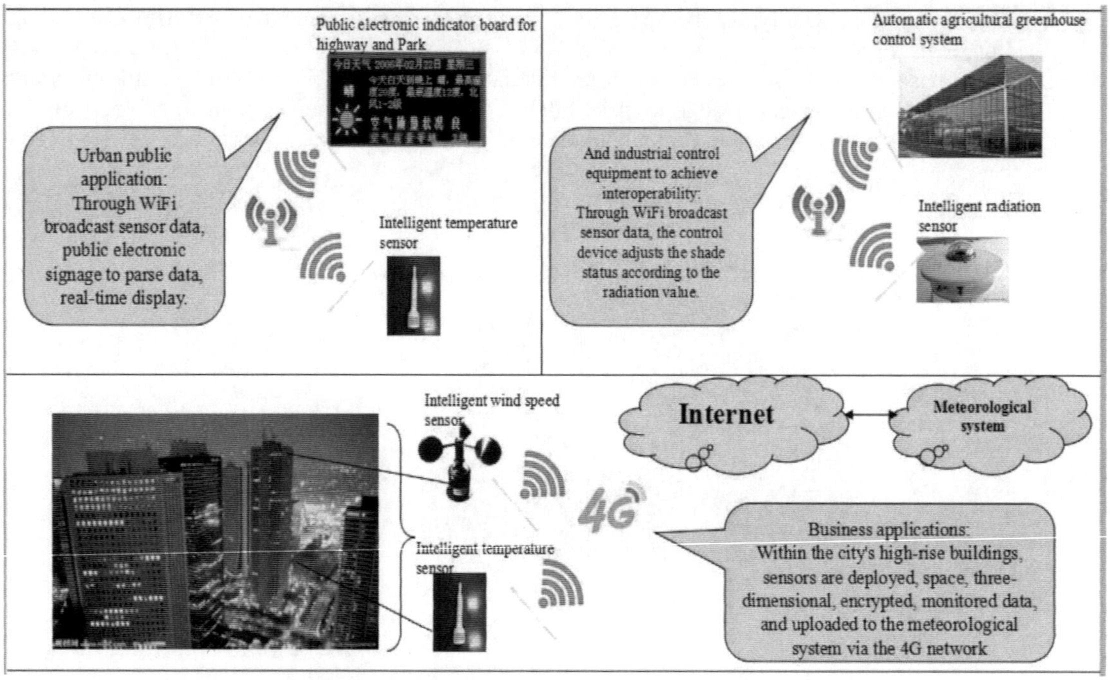

Figure 4.shows intelligent sensor independent deployment mode.

4. Building of the IoT automatic station

4.1. Construction object

1) Should be based on a small Internet data automatic transmission station, massive data collection and analysis, to achieve intelligent temperature and humidity sensor and rain sensor independent work and send data through the NB-iot communication mode, reduce power consumption, extend equipment life cycle. In the cloud to achieve data integration, and through intelligent data analysis algorithm to determine the sensor running state, to enhance the ability of meteorological data interaction and information technology level.
2) Intelligent sensor independent work, data transmission wireless. Research on networking data collection and dissemination of meteorological data, through the data platform for data analysis, the preliminary work of meteorological information publishing standards, networking of meteorological information receiving terminal provides the data interface, to the wisdom of the city, the wisdom of the purpose of the meteorological service.

4.2. The contents of the building

1) Build the IoT automatic station including conventional six elements (temperature, humidity, pressure, rain, wind direction and wind speed) based on digital sensors and Zigbee wireless network technology.
2) Find the deficiency of existing equipment through the trial run. Enhance the intelligent level of the sensor and the overall reliability of the Internet of things comprehensively through the diversity of observation, the existing data fusion between sensors and integrated quality control and other technical means.
3) Finally explore a kind of IoT technology that suitable for automatic weather station mostly: the next generation of IoT technology represented by NBiot is different from the Zigbee, wireless network current Internet of things model, try to build Internet of things automatic station based on NBiot.

References

[1] Zhipengzhao, lianmeishi. Research on meteorological disaster monitoring and early warning system based on Internet of things technology.Computer knowledge and technology. 2016， 12（08）

[2] Li Song, Li Yong, Zhang Chao, Xia Wei.Application of Internet of things technology in meteorological service.SME management and technology.2016(01):223

[3] Li Jianbang, Zhou Shuxue, Li Aihua, Yuan Ye.Application of Internet of things in weather modification service in Anhui Province.Meteorological science and technology.2014，42(6):1143-1146.

[4] zhongyong.Application of Internet of things in weather disaster early warning.Communication and information technology.2012(02):78-79

[5] rongfang yang. Application of Internet of things in meteorological disaster prevention and mitigation. Atmospheric composition and climate change.2012(S16)

Residents' Perceptions of Walkability Attributes in Mainland China: Reliability and Validity

Yang Ye[1] **, Teng Fei**[2] **and Hongyuan Mei**[1, 2]

[1]School of Architecture, Harbin Institute of Technology, No.66 Xidazhi Street, Harbin, China
[2]Architecture Design and Research Institute of Harbin Institute of Technology, No.93 Huanghe Road, Harbin, China

*Corresponding Author: 191775660@qq.com

Abstract. To test the influence of urban built environment on walkability in mainland Chinese cities, this study developed a questionnaire to perceive the residents' perception of walkability around their neighborhood, which was based on the Neighborhood Environment Walkability Scale-Abbreviated (NEWS-A), and modified to reflect the characteristics of mainland Chinese built environment and people's behavior mode. To perceive the final version of Neighborhood Environment Walkability Scale for Mainland China (NEWS-MC), We made a pilot study which include 50 samples, then interview 350 residents from 4 selected residential districts in Harbin which were different in walkability attributes and economic status. The final version of the NEWS-MC included 8 subscales and 5 single items (81 items in total). Test-retest reliability showed moderate to high except 1 subscale. In total, the NEWS-MC could illustrate residents' perceptions of walkability attributes in mainland Chinese cities and could be use in other Chinese urban attributes studies related to walking.

1. Introduction

The high speed of cities sprawl and urban regeneration worldwide make public space and acclivities in public space gain increasingly attention than ever. Meanwhile, there is growing interest on research about relationship between built environment attributes and residents' physical acticity [1], [2], [3], [4]. Studies focus on walking suggest than higher residential density, higher level of land use mix- both diversity and accessibility, more aesthetic elements are attributes could hence more walking, and associated with neighborhoods more walkable[5], [6].

'Walkability' is referred to and generally defined as "...the extent to which walking is readily available as a safe, connected, accessible and pleasant mode of transportation". In the last decade, there were increasing evidence of the influence of neighborhoods built environment characteristics on walkability. Smith and Clifton illustrated that an built environment friendly to pedestrians usually includes characteristics such as high density, convenience to walk to facilities, system coherence, high street connectivity, safety from traffic and crime, comfortable and attractiveness[7]. Researcher in urban planning and transportation also identified key environment attributes on creating walkable neighborhood. Cervero and Duncan found density, trip distance, land use diversity, existing of sidewalk and bike lane could affect people's choice of walk or bike [8]. The studies by De bourdeaudhuij demonstrated the effects of space proportion, size of block and traffic congestion on user's choice of travel mode [9].

To make more research on influence of urban environment's walkability on physical activities and psychological health or residents, valid measures of walkability are needed. The Neighborhood Environment Walkability Scale(NEWS) and its abbreviated version(NEWS-A) are developed for use in USA to measure residents' perceptions of the environment attributes[10]. They value built environment characteristics that are believed to affect residents' walking choice. Both NEWS and NEWS-A have been found have moderate to high rest-retest reliabilities[11], and good criterion The evidence illustrated high-walkable neighborhoods be found with higher density , land-use mix, street connectivity, more safety and have more aesthetics elements[12]. Although NEWS and NEWS-A have been applied in many countries, such as US, Australia, Korea and Hong Kong, the validity and reliability of the NEWS or NEWS –A have not been examined in mainland China.

2. Method

2.1. Measures

Neighborhood Environment Walkability Scale(NEWS) [13] is a instrument used to obtain residents' perception of built environment attributes related to walking and bicycling around where they live. The original NEWS includes 98 items and is rarely used, the abbreviated version of it, NEWS-A [14], which has similar structure to the full version but has fewer items in subscales, is commonly used and have been applied in many countries (Brazil, Australia, Poland, Korea, Mexico, and Hong Kong). So in this study we also used NEWS-A as the source to develop the NEWS version which appropriate for Mainland China (NEWS-MC). NEWS-A has 8 multi-item subscales, a) residential density; b) land-use mix-diversity; c) Land-use mix-access; d) street connectivity; e) infrastructure and safety for walking; f) aesthetics; g) traffic hazards and h) Crime safety, and 5 single-item subscales, including Lack of parking, lack of cul-de-sacs, hilliness, physical barriers and social interaction while walking. All items except the ones in residential density and land-use mix-diversity subscales are rated on a 4-point Likert scale (1- strongly disagree, 2-disagree, 3-agree, 4-strongly agree). Items in residential density subscale are rated on 5-point Likeert scale, by asking about the frequency of perceive certain type of residences, and weighted ratings are summed to make a residential density perceptions score. Land-use mix-diversity is assessed by time walking from home to various types of destinations, with responses ranging from distance for 1-5mins walking (coded as 5) to distance for more than 30mins walking (coded as 1).

2.2. Participants and procedures

2.2.1 Development and adaptation The original NEWS and NEWS-A were developed in US, to develop a version of NEWS which could adapt Chinese cities' build environment attributes and reflects residents' need, 15 experts from multi-research fields, urban planning and urban design(7members), public health(3 members), environment-behavior studies(2 members), physical activity(2 members) and Rehabilitation & Healthcare(1 member) were interviewed with a goal to draft a NEWS-MC. All memebers of the committee reviewed NEWS-A, we also provided them NEWS-Youth[15], NEWS for senior citizen[16] and NEWS Hong Kong[17] as reference. Committee members gave their opinions on whether each item should be kept, modified or deleted, or any item should be add after careful thinking about items relevant to mainland Chinese urban environment characteristics and residents' behavior mode. In total, 57 items in NEWS-A were kept, 9 of them were modified, 17 new items were added and 1 item was deleted considering their relevant to the study setting. (Table 1).

Items in residential density subscale were modified to fit the medium to high residential density characteristic in mainland Chinese cities. 'detached single-family residences' and 'townhouses or row hoses of 1-3 stories' were merged into a single item as these kind of residential buildings are very rarely in mainland china; apartments of 4-6, 7-12 and more than 13 stories were modified to apartment of 4-7, 8-18, 19-26 stories and apartments of 27 stories or more was added to this subscale, as 7, 18, 26, 27 and above stories residential buildings are the extreme height for multi-stories, mid-high rise, high rise and super high rise residential buildings through the Chinese building codes. 5 items in land –use mix-diversity were also modified to fit Chinese urban environment(see Table 1). 17 items were added to the questionnaire to provide a better match mainland Chinese urban built environment features and culture background- medium to high population density and building density, high percentage of walking or using public transportation in daily travel, high risk of air pollution and safety issues related by traffic and crime. Item 'video store' was deleted because it is very rarely in mainland Chinese cities now, maybe caused by the development of internet everyone watch or buy movie or video online(see Table 1).

2.2.2 Test-retest reliability. Potential participants were chosen from the four Harbin residential neighborhoods mention above, the economic status were obtained from national 2015 Census data. Walkability was measured by residential density, intersection density, destination density (service, commercial and facilities), road density(length of street could walk per km^2) and type of street block(square grid, radial shape or have many cul-de-sacs). These data were achieved from baidu map (http://map.baidu.com/).

Four residential neighborhoods, with 125 potential participants were chosen randomly from each neighborhood. The survey were done in March 2017, We went to each neighborhood in both weekday and weekend, morning, noon and evening. When picking the participants, firstly we found the people showed up in that area and asked whether they lived in this neighborhood, if the answer was yes, we further asked whether they would be willing to be reassessed 10 days later after the first survey, and they were told if they finished both survey, they will get 10RMB as reward. After confirm lived in that residential neighborhood and agreed to attend both survey, one became our participant. The process was the interviewer and the participants added each other as 'friend' on wechat (a social App with high popularity in smart phone holders of all ages, add as friends mean two can message each other, and could also pay and receive money, receive and send documents to each other) , then sent them the NEWS-MC questionnaire forms, the participants could contacted the interviewer by wechat if they had any questions about the questionnaire. After they sent the first-time forms back, they could get 3RMB. 10 days later we sent them the forms again, and gave who sent forms back 7RMB, the ones who did not sent form back were reminded twice on wechat. The interval between the two surveys ranged from 10-16 days(average 12.4 days) . The recruitment rate of the first survey was 85%, 92, 93, 85 80and participants responded the second survey, so the total recruitment rate is 70%.

Table 1. Adaptations of the NEWS-A for Mainland Chinese

Item
Addition items
Distance to community doctor service
Distance to Chinese drink or noodle shop
Distance to Chinese non-fast food restaurant
Distance to Bakery/dessert/cake shop
Distance to community centre
Distance to copy /scan passport photo-taken shop
Distance to place could pay all kinds of fees(gas, water, cable TV and internet)
Distance to pets' food or clothes shop
Shopping areas are easily accessible via public transport.
The streets are so crowded that it is difficult to walk.
Need to walk over a pedestrian bridge or through a wide road to access the nearest services
There are 'hawkers' and shops on the streets and sidewalks blocking the way
There are many cars parking on the sidewalks blocking the way
There are indoor, air-conditioned places (shopping malls) where people can walk
The streets of my neighborhood are often slippery in winter
The level of air pollution in my neighborhood is often high
It is unsafe to walk in my neighborhood because of objects dropping from high-rise buildings

There are many homeless people, drunk people in my neighborhood
Modifications items
Two items detached single-family residences and townhouses or row hoses of 1-3 stories merged into a single item apartments of 4-6 stories change to apartment of 4-7 stories. apartments of 7-12 stories change to apartment of 8-18 stories. apartments of more than 13 stories changed to 'apartments of 18-26' stories. Additional item referring to commonness of apartment buildings with more than 26 stories added Destination 'fruit/vegetable market' renamed 'fresh food market' Destinations 'fast-food restaurant' renamed 'Chained Western or Chinese fast-food restaurant' Destination 'non-fast food restaurant' changed to 'Western/international non-fast food restaurant' Destination 'recreation centre' changed to 'recreation centre or outdoor recreation space' Destination 'clothing store' changed to 'clothing & shoe store' 'Walkers and bikers on the streets can be easily seen by people in their homes' changed to 'walkers and bikers on the streets can be easily seen by other people'
Deleted items
Destination 'video Store'

3. Results

Four hundreds and twenty-five participants attended the first survey, with 227 men and 198 woman, mean age 45.2. 48% participants had a collage degree or above. Among all participants, 32% of them own 1 car or more, and 45% reported they use public transportation 'very often'. Three hundreds and fifty participants completed both surveys (193 men and 157 woman), response rate was 82%. The sample size per neighborhood were: 90 and 93 in high-walkable neighborhoods, 82 and 85 in low-walkable neighborhoods.

Through previously built satisfaction systems, we took ICC value below 0.50 as poor, 0.50 to 0.75 as moderate, and above 0.75 as reliability. Test-retest reliability for all subscales and single items of the NEWS-MC are shown in Table 2. All subscals and single-items except one-hilly roads-showed moderate to good level of reliability(>0.75), 6 subscales or single items show high level of reliability (>0.90), and all p<0.05.

Table 2 illustrates the results of mean scores on neighborhood walkability subscales between residents from high-walkable and low-walkable neighborhood. Residents from high-walkable neighborhoods provided ratings indicative of higher residential density(t(412)= 6.85, p<0.001), land-use mix-diversity(t(420)= -10.97, p<0.001); Land-use mix-access (t(420)=4.15 , p<0.001); street connectivity(t(415)= 2.93, p<0.001), infrastructure and safety for walking(t(397)= 5.32 , p=0.006), traffic hazards(t(399)=2.07 , p=0.021), parking difficulties(t(408)= 6.81, p=0.004), physical barriers(t(388)= 8.74, p<0.001) and social interaction while walking(t(386)= -4.85, p<0.001). Mean while, residents from low-walkable provided higher ratings of aesthetics(t(418)= -6.28, p=0.008) of their neighborhoods. Residents' perception of crime safety(t(385)= 8.76, p<0.315), not have many cul-de-sacs(t(408)= 5.79, p<0.438), and hilliness(t(408)= 5.79, p<0.570) did not show difference between low and high walkable neighborhoods.

4. Discussion

The result of this study indicated the link between walkability and residents' perception of built environment attributes. By using NEWS-MC, a form designed to obtain residents' perception of walkability attributes in medium to high density mainland Chinese cities, as instrument, this study examined the test -retest the reliability and validity of the subscales and single items of NEWS-MC. Moderate to high level of reliability was founded from the result of test-retest reliability of NEWS-MC. The intra-class correlation range from 0.65 to 0.98, which showed a higher reliability than the result in United States(0.58-0.80), Australia(0,62-0,88) and comparable to those observed in Hong Kong(0.57-0.99). The difference may because in US and Australia the researcher used self-administered survey, in this study although we sent the NEWS-MC form to participates but they could communicate with the interviewer by wechat so it kind of like a interviewer administered survey which was used in Hong Kong study [17]. Besides, the time interval of first and second survey was 10-16 days, which is shorter than the previous studies (14-21 days).

The result of this study support the applicability of the NEWS-MC to medium to high density cities in mainland China and most Asian cities, such as Harbin. The data showed significant difference in residents' perception on built environment characteristics between high-walkable and low-walkable urban areas. Residential density, land-use mix-diversity, Land-use mix-access, street connectivity, infrastructure and safety for walking, traffic hazards, parking difficulties, physical barriers and social interaction while walking illustrated residents could perceive higher level of attributes in these subscales from high-walkable neighborhoods. The walkability of the neighborhoods were objectively measured by residential density, land-use mix, and pedestrian sidewalk connectivity, and the residents self-reported result reflected the difference of perceiving these attributes. It is observed that the different direction of ratings on aesthetics, low-walkable neighborhoods were rated higher in most of the items in this subscale, which is the same in Australia and Hong Kong, may be attributed to residents' preference open view and street streets and more green bushes or grass which the low-walkable neighborhoods had more to other aesthetic elements, such as street furniture and sculpture, which the high-walkable neighborhoods had more. The insignificant on crime, hilliness and cul-de-sacs subscales were likely because Harbin is a flat city without many hill and the public order is very good here, and the road network in Harbin is basically grid or radiate, the cul-de-sacs is very rare even in residential area, so residents can not perceive the relationship between items in these subscales or single items and walking.

Table 2. NEWS-mainland China Test-retest reliability and mean subscale scores for neighborhoods with different walkability.

Subscale/single item(no. Items)	Test-retest reliability(N=350)	High-walkable (N=183)Mean(SD)	Low-walkable (N=167)Mean(SD)
Residential density(6)	0.95	3.26(0.27) [a]***	2.58(0.31)
Land-use mix – diversity[c](28)	0.88	4.24(0.40) [a]***	2.48(0.28)
Land-use mix – access(5)	0.93	3.82(0.32) [a]***	1.65(0.42)
Street connectivity(3)	0.81	3.59(0.33) [a]***	2.38(0.42)
Infrastructure and safety for walking(11)	0.97	2.48(0.39) [a]**	2.12(0.38)
Aesthetics(6)	0.79	2.42(0.58) [b]**	3.15(0.42)
Traffic and road hazards[c](13)	0.86	3.20(0.30) [a]*	3.01(0.21)
Crime[c](5)	0.98	2.41(0.41)	2.31(0.37)
Parking difficulties(single item)	0.88	2.17(0.55) [a]**	1.92(0.32)
Not many cul-de-sacs(single item)	0.92	3.24(0.45)	3.25(0.51)
Hilly roads[c](single item)	0.65	1.85(0.23)	1.76(0.28)
Physical barries[c] (single item)	0.91	2.86(0.44) [a]***	2.42(0.38)
Interactive to other people(single item)	0.94	3.58(0.31) [a]***	2.27(0.39)

Note: ICC = intraclass correlation coefficient

[a] High walkable>Low walkable, [b] Low walkable>high walkable

[c] this subscale reverse scored to reflect the same direction of other environment characteristics, higher scores means higher walkabiltiy.

*p<0.05, **p<0.01, ***p<0.001

The mean values comparison between NEWS-MC to NEWS Untied States, Australia and Hong Kong indicated that, as hypothesis, residential density was rated much higher than US and Australia but lower than Hong kong. This can be attributed to the fact that the overall mainland Chinese cities are medium to high density, the urban residential density is higher than US and Australia, but lower than Hong Kong. Land-mix diversity was rated higher in both high-walkable and low-walkable neighborhoods than all the other three studies. It should be notice that the previous studies all used the long version NEWS and we used NEW-A the measure instrument, precluded a direct comparison of the findings.

The limitations of this study include:1) The participants are the ones who will to attend the survey, not matched on individual respondent demographic. 2) we only do research in for residential neighborhoods, the small number of areas and small sample number of participants may have limited the reliability and variability of some items. 3) We may omitted some environment attributes that related to walkability of medium to high density mainland Chinese cities.

In conclusion, the result of this study support reliability and construct validity of the NEWS-MC, so we could take the NEWS-MC as a instrument for further study of cross-national or international research on the relationship between built environment characteristics and residents' walking choice.

Acknowledgments

This study was supported by Provincial Natural Science Founding of Heilongjiang (JJ2017QN0308); Harbin Application Technology Research and Develop Project Science and Technology Innovation Talented Person (2016RAYBJ001).

References

[1] Humpel, N., Owen, N., & Leslie, E. (2002). Environmental factors associated with adults' participation in physical activity: a review.*American Journal of Preventive Medicine, 22*(3), 188-199.

[2] Killingsworth, R. E. (2003). Health promoting community design: a new paradigm to promote healthy and active communities. *American Journal of Health Promotion Ajhp, 17*(3), 169.

[3] Green, W. A., & Stoltz, R. R. (2004). Health and community design: the impact of the built environment on physical activity by lawrence d. frank, peter engelke and thomas schmid. *Landscape & Urban Planning, 69*(4), 467-470.

[4] Sallis, J. F., Bauman, A., & Pratt, M. (1998). Environmental and policy interventions to promote physical activity. *American Journal of Preventive Medicine, 15*(4), 379-397.

[5] Lawrence D. Frank, James F. Sallis, Terry L. Conway, James E. Chapman, Brian E. Saelens, & William Bachman. (2006). Many pathways from land use to health: associations between neighborhood walkability and active transportation, body mass index, and air quality. *Journal of the American Planning Association, 72*(1), 75-87.

[6] Krizek, K. J. (2000). Pretest-posttest strategy for researching neighborhood-scale urban form and travel behavior. *Transportation Research Record Journal of the Transportation Research Board,1722* (1722), 48-55.

[7] Clifton, K. J., Smith, A. D. L., & Rodriguez, D. 2007. The development and testing of an audit for the pedestrian environment. *Landscape & Urban Planning, 80*(1–2), 95-110.

[8] Cervero, R., & Duncan, M. 2011. Walking, bicycling, and urban landscapes: evidence from the san franci. *American Journal of Public Health, 93*(9), 1478-1483.

[9] De Boudeaudhuij, I., Sallis, J. F., Saelen, B. E., De Bourdeaudhuij, I., Saelens, B., & De Bourdeaughuij, I., et al. 2003. Environmental correlates of physical activity in a sample of berlgian adults. *American Journal of Health Promotion, 18*(1), 83-92.

[10] Cerin, E., Conway, T. L., Saelens, B. E., Frank, L. D., & Sallis, J. F. 2009. Cross-validation of the factorial structure of the neighborhood environment walkability scale (news) and its abbreviated form (news-a).*International Journal of Behavioral Nutrition and Physical Activity, 6*(1), 1-10.

[11] Leslie, E., Saelens, B., Frank, L., Owen, N., Bauman, A., & Coffee, N., et al. 2005. Residents' perceptions of walkability attributes in objectively different neighbourhoods: a pilot study. *Health & Place, 11*(3), 227-236.

[12] Saelens, B.E., Sallis, J.F., Frank, L.D., 2003b. Environmental correlates of walking and cycling: Findings from the transportation, urban design, and planning literatures. *Annals of Behavioral Medicine 25* (2), 80–91.

[13] Saelens, B. E., Sallis, J. F., Black, J. B., & Chen, D. 2003. Neighborhood-based differences in physical activity: an environment scale evaluation. *American Journal of Public Health, 93*(9), 1552.

[14] Cerin, E., Leslie, E., Owen, N., & Bauman, A. 2008. An australian version of the neighborhood environment walkability scale: validity evidence. *Measurement in Physical Education & Exercise Science,12*(1), 31-51.

[15] Rosenberg, D., Ding, D., Sallis, J. F., Kerr, J., Norman, G. J., & Durant, N., et al. 2009. Neighborhood environment walkability scale for youth (news-y): reliability and relationship with physical activity. *Preventive Medicine, 49*(2–3), 213–218.

[16] Shigematsu, R., Sallis, J. F., Conway, T. L., Saelens, B. E., Frank, L. D., & Cain, K. L., et al. 2009. Age differences in the relation of perceived neighborhood environment to walking. *Medicine & Science in Sports & Exercise, 41*(2), 314.

[17] Cerin, E., Macfarlane, D. J., Ko, H. H., & Chan, K. C. A. 2007. Measuring perceived neighbourhood walkability in hong kong. *Cities, 24*(3), 209-217.

Empirical Study on Total Factor Productive Energy Efficiency in Beijing-Tianjin-Hebei Region-Analysis based on Malmquist Index and Window Model

Qiang Xu[1, 2], Shuai Ding[1, *] and Jingwen An[1]

[1]School of Management, China University of Mining and Technology, Beijing 100083, China
[2]Beijing Municipal Science & Technology Commission, Beijing 100195, China

[*]corresponding author e-mail:dsahau@yeah.net

Abstract. This paper studies the energy efficiency of Beijing-Tianjin-Hebei region and to finds out the trend of energy efficiency in order to improve the economic development quality of Beijing-Tianjin-Hebei region. Based on Malmquist index and window analysis model, this paper estimates the total factor energy efficiency in Beijing-Tianjin-Hebei region empirically by using panel data in this region from 1991 to 2014, and provides the corresponding political recommendations. The empirical result shows that, the total factor energy efficiency in Beijing-Tianjin-Hebei region increased from 1991 to 2014, mainly relies on advances in energy technology or innovation, and obvious regional differences in energy efficiency to exist. Throughout the window period of 24 years, the regional differences of energy efficiency in Beijing-Tianjin-Hebei region shrank. There has been significant convergent trend in energy efficiency after 2000, mainly depends on the diffusion and spillover of energy technologies.

1. Introduction

Energy is the fundamental driving force of the world economy and development and is the material basis for human survival. Under the rapid development of the whole world, the development of energy and the protection of environment are the focus of mutual attention of the whole world and of all mankind, this is also one of the most important issues facing China's transformation of economic development mode. According to the 2012 "BP world energy statistics yearbook", in 2010, China's energy consumption per unit of GDP was 0.63 tons of standard oil/thousand dollars, it's 2.5 times the world average. From this index, Italy has the highest energy efficiency (lowest energy consumption intensity), only 0.18 tons of standard oil/thousand dollars. That means, to create the same economic benefits, Italy's energy consumption is about a third of our energy consumption. The 13th five-year plan refers to the major goals and basic concepts of economic and social development, in particular, it emphasizes "the development and utilization efficiency of energy resources has been greatly improved. Energy and water consumption, construction land and total carbon emissions are effectively controlled". Therefore, to study the energy efficiency of Beijing-Tianjin-Hebei region, and find the change in energy efficiency, has practical significance to improve the economic development quality in Beijing-Tianjin-Hebei region.

2. Literature review

The existing literature of research on energy efficiency is quite fruitful. Qu Xiaoe (2009) empirically measured the total factor energy efficiency and technological progress and technical efficiency index in 30 provinces in China from 1990 to 2006. The conclusion shows that, structural adjustment, technological progress and energy price increase have positive effects on energy efficiency improvement [1]. Wang Qunwei et al. (2010) studied the spatial and temporal variation characteristics of energy utilization level in China through the total factor energy efficiency, research shows that, the difference of China's regional energy efficiency is remarkable, technological progress, pure technical efficiency and scale efficiency are of great importance to the improvement of energy efficiency [2]. Li Lanbing (2012) based on the dual view of management and environment, applied the DEA four-stage model to analyze the energy efficiency of China's provincial-level regions from 2005 to 2009. The conclusion shows that, the total factor energy efficiency in China is still at a low level in general [3]. Cheng Yuhong and Zhang Weiqi (2013) used panel data from 21 cities in Guangdong province from 2005 to 2010, empirically analyzed the characteristics of the total factor energy efficiency, research shows that, there are obvious regional differences in total factor energy efficiency, the urban economic development level has uncertainty relationship on the total factor energy efficiency [4]. Yang Lili et al. (2014) used the stochastic frontier analysis method to measure the industrial total factor energy efficiency growth rate in the 14 representative cities of 2001 to 2009. The conclusion shows that, the scale of enterprises, opening to the outside world, government intervention, foreign direct investment and the proportion of coal consumption have a significant inhibitory effect on the industrial total factor energy efficiency growth [5]. Chen Guanjv (2014) used random frontier production technology to measure the total factor energy efficiency of 30 industries of China's manufacturing industry, research shows that, the energy efficiency level between industries is quite different [6].

Based on the above research results, in consideration of the difference in energy efficiency in Beijing-Tianjin-Hebei region, it is not appropriate to assume uniform production functions. Therefore, the paper uses the Malmquist index and Window Analysis (WA) model method based on DEA, uses the panel data from 1991 to 2014, changes and decomposes energy efficiency in Beijing-Tianjin-Hebei region, and finds trends in energy efficiency in order to obtain some valuable conclusions and to give corresponding policy suggestions.

3. Research methods, variables and data processing

3.1. Research method

The Malmquist index method is defined based on the distance function. The linear optimization method is used to measure the efficiency change and technological progress. The change of Malmquist index is the change of total factor productivity (TFP). Under time t's technical conditions, the Malmquist index based on the output Angle can be expressed as:

$$M_o^t = D_o^t\left(x^{t+1}, y^{t+1}\right) / D_o^t\left(x^t, y^t\right) \tag{1}$$

In formula (1), $D_o{}^t$ is distance function, the subscript $_o$ represents the distance function based on output, x and y are inputs and outputs. In the same way, under time (t+1)'s technical conditions, the Malmquist index based on the output Angle can be expressed as:

$$M_o^{t+1} = D_o^{t+1}\left(x^{t+1}, y^{t+1}\right) / D_o^{t+1}\left(x^t, y^t\right) \tag{2}$$

When the index is greater than 1, the total factor productivity of the period from t to (t+1) is increasing, as shown in formula (3):

$$M_o(x^{t+1}, y^{t+1}, x^t, y^t) = \left[\frac{D_o^t(x^{t+1}, y^{t+1})}{D_o^t(x^t, y^t)} \times \frac{D_o^{t+1}(x^{t+1}, y^{t+1})}{D_o^{t+1}(x^t, y^t)}\right]^{1/2} = \frac{D_o^{t+1}(x^{t+1}, y^{t+1})}{D_o^t(x^t, y^t)} \times \left[\frac{D_o^t(x^{t+1}, y^{t+1})}{D_o^{t+1}(x^{t+1}, y^{t+1})} \times \frac{D_o^t(x^t, y^t)}{D_o^{t+1}(x^t, y^t)}\right]^{1/2} = TE \times TP = PE \times SE \times TP = TFP \quad (3)$$

The window model is a dynamic efficiency analysis model for panel data based on the traditional DEA model. Window analysis is operated under the principle of moving average. First, select the window width d of the study, Charnes et al. (1994) think, choosing window width d = 3 or d = 4 can achieve the best balance in the stability and reliability of the efficiency measure [7]. The window model can dynamically evaluate the relative efficiency of a decision unit, monitor its performance, and provide a basis for managers to make effective decisions [8].

3.2. Variable selection and data processing

Based on traditional economic growth theory, the article selects capital stock (K), human capital stock (L) and total energy consumption (E) as input variables, selects the gross domestic product (GDP) as the output variable, measured and analyzed the total factor energy efficiency in Beijing-Tianjin-Hebei region from 1991 to 2014. Data is from "China energy statistical yearbook", "China statistical yearbook", "Beijing statistical yearbook", "Tianjin statistical yearbook", "Hebei economy yearbook" and "China statistical data collection" in related years. The variable indexes of input and output are selected as follows:

1) Capital stock (K). According to the existing literature, in this paper, the capital stock of Beijing-Tianjin-Hebei region is estimated by using the popular "perpetual inventory method" (Goldsmith, 1951), which is widely adopted by scholars. The calculation method is: $K_{i,t}=I_{i,t}+(1-\sigma_i)\ K_{i,t+1}$. Among them, $K_{i,t}$ is the capital stock of t years in i region. $I_{i,t}$ is the investment of t years in i region. σ_i is the depreciation rate of fixed assets in i region. This paper is based on 1990 price base period, and uses the regional fixed asset price index to reduce, removes price factor fluctuations; uses regional fixed capital formation to represent investments $I_{i,t}$, fixed assets depreciation rate σ_i uses 9.6%.
2) Human capital stock (L). From the previous literature, the researchers use average educational year index to measure human capital stock. It has certain representativeness. Therefore, this paper chooses "average educational year" to reflect the human capital stock in Beijing-Tianjin-Hebei region. Among them, the university graduation education years are 16 years, high school graduation are 12 years, junior high school graduation are 9 years, primary school graduation are 6 years.
3) Total energy consumption (E). This paper uses the total energy consumption in Beijing-Tianjin-Hebei region.
4) The gross domestic product (GDP). This paper is based on the real GDP of Beijing-Tianjin-Hebei region, reduces the price factor by reducing the price of GDP which is based on 1990 price base period.

4. The empirical analysis

4.1. DEA-Malmquist analysis

In order to explore the intrinsic factor of the total factor energy efficiency in Beijing-Tianjin-Hebei region, this article uses the DEAP 2.1 software package to estimate the total factor energy efficiency (TFEE) in Beijing-Tianjin-Hebei region from 1991 to 2014, and decomposes it into the energy efficiency change index (TEE) and the energy technology progress index (TPE), TEE can be decomposed into the energy pure technology efficiency change index (PEE) and energy scale efficiency change index (SEE). The results are shown in table 1.

Table 1. Change and decomposition of total energy efficiency in Beijing-Tianjin-Hebei region

YEAR	TFEE	TEE	TPE	PEE	SEE
1991-1992	1.036	0.987	1.049	1.000	0.987
1992-1993	1.023	1.005	1.018	1.000	1.005
1993-1994	1.045	1.005	1.040	1.000	1.005
1994-1995	1.016	1.000	1.016	1.000	1.000
1995-1996	1.038	1.011	1.027	1.000	1.011
1996-1997	1.048	1.000	1.048	1.000	1.000
1997-1998	1.023	1.000	1.023	1.000	1.000
1998-1999	1.021	1.000	1.021	1.000	1.000
1999-2000	0.985	1.000	0.985	1.000	1.000
2000-2001	1.017	1.000	1.017	1.000	1.000
2001-2002	1.013	1.000	1.013	1.000	1.000
2002-2003	1.015	1.000	1.015	1.000	1.000
2003-2004	0.997	1.000	0.997	1.000	1.000
2004-2005	1.012	1.000	1.012	1.000	1.000
2005-2006	1.020	1.000	1.020	1.000	1.000
2006-2007	1.010	1.000	1.010	1.000	1.000
2007-2008	1.031	1.000	1.031	1.000	1.000
2008-2009	1.022	1.000	1.022	1.000	1.000
2009-2010	1.031	1.000	1.031	1.000	1.000
2010-2011	1.006	1.000	1.006	1.000	1.000
2011-2012	1.004	1.000	1.004	1.000	1.000
2012-2013	1.043	1.000	1.043	1.000	1.000
2013-2014	0.992	1.000	0.992	1.000	1.000
MEAN	1.019	1.000	1.019	1.000	1.000

In terms of time, the average total factor energy efficiency index of Beijing-Tianjin-Hebei region from 1991 to 2014 was 1.019, it has already on the frontier of energy efficiency, its average growth rate is 1.9 %(The total factor energy efficiency index minus 1 is the growth rate (Fare, etc., 1994) [9]), so the level of total factor energy efficiency is rising. We can see from the further breakdown of the TFEE index, the average growth rate for TPE in the same period was 1.9%, TEE didn't change. Therefore, the total factor energy efficiency of Beijing-Tianjin-Hebei region depends mainly on technological progress or innovation.

From the regional dimension, as shown in table 2, the highest total energy efficiency index of Beijing was 1.029, its growth rate was 2.9%. Tianjin was second to 1.015, its growth rate was 1.5%. Hebei is 1.014, but it also grew by 1.4%. It can be seen that, the increase and decrease of total factor energy efficiency in both cities and provinces is mainly due to technological progress. Therefore, accelerating the technological progress and enhancing the capacity of independent innovation is the main way to improve the total factor energy efficiency in Beijing-Tianjin-Hebei region.

Table 2. The average change and decomposition of total factor energy efficiency in Beijing-Tianjin-Hebei region

REGION	TFEE	TEE	TPE	PEE	SEE
Beijing	1.029	1.000	1.029	1.000	1.000
Tianjin	1.015	1.001	1.014	1.000	1.001
Hebei	1.014	1.000	1.014	1.000	1.000
Beijing-Tianjin-Hebei	1.019	1.000	1.019	1.000	1.000

4.2. DEA window analysis

In order to further analyze the differences and trends of energy efficiency in Beijing-Tianjin-Hebei region, this article uses the DEA - Solver Pro5.0 software package, selecting window width d = 3, that is to say 3 years to make up a time window, under BCC model based on input angle, and calculates the average efficiency of the windows energy efficiency in Beijing-Tianjin-Hebei region.

Overall, the energy efficiency of Beijing-Tianjin-Hebei region shows obvious differences in different window periods. Specific, the regions with the highest energy efficiency in the window 1-10 are all Hebei, the average efficiency of windows 1, 2, 5, 6 and 7 has reached 1. The highest average efficiency of window 11 is Beijing (0.9994). The highest average efficiency of window 12-13 is Hebei. The efficiency values are 0.9994 and 0.9998 respectively. The highest average efficiency of window 14 is Tianjin (0.9998). The highest average efficiency of window 15 is Tianjin and Hebei, the efficiency value is 1. The highest average efficiency of window 16-17 is Tianjin, the efficiency values are 0.9995 and 1 respectively. The highest average efficiency of window 18 is Beijing (1). The highest average efficiency of window 19 is Hebei (0.9967). The highest average efficiency of windows 20 is Tianjin and Hebei, the efficiency value is 1. The highest average efficiency of window 21-22 is Tianjin, the efficiency value is 1.

From the extreme difference, the difference in energy efficiency in Beijing-Tianjin-Hebei region can be reflected from the side of extreme difference of window 1-22. This difference rose from 0.0285 of window 1 to 0.0356 and 0.0296 of windows 6 and 7. They increased by 24.91% and 3.86% respectively. It then drops to 0.0009 of the lowest window 15, the drop was as much as 97.47 percent. Then rise to 0.0016 of window 22, the rise is 77.78%. The overall change in extreme difference was -94.39%. This indicates that the energy efficiency gap among Beijing-Tianjin-Hebei region has been shrinking over the past 24 years.

From the window changes, during 24 years from 1991 to 2014, energy efficiency in Beijing-Tianjin-Hebei region has generally increased, and the mean has increased by 1.05%. The region where energy efficiency has increased most is Beijing, the growth is 2.77%. Next is Tianjin, the growth is 0.54%. But there has been a marked decline in Hebei, the drop is 0.10%. The energy efficiency of Beijing-Tianjin-Hebei region has shown a clear convergence trend since 2000.

5. Conclusions and policy recommendations

5.1. Main conclusions

This paper combines DEA - Malmquist and window analysis model, uses panel data from 1991 to 2014 in Beijing-Tianjin-Hebei region, empirically measures the total factor energy efficiency in Beijing-Tianjin-Hebei region, and carries out the decomposition of the change of total factor energy efficiency, discovers the trend of energy efficiency. The following main conclusions are drawn.

The total factor energy efficiency of Beijing-Tianjin-Hebei region is rising, its average growth rate is 1.9%. It relies mainly on energy technology progress or innovation. From 1991 to 2014, there are regional differences in total energy efficiency in Beijing-Tianjin-Hebei region and it appears fluctuating trend of "first rise and then rise". Among them, the highest total energy efficiency index is Beijing and it is 1.029, its growth rate was 2.9%. Tianjin was second to 1.015, Hebei is 1.014. All the total factor energy efficiency of Beijing-Tianjin-Hebei region has reached the frontier of energy efficiency, and the change is mainly due to technological progress.

There is a convergence trend in energy efficiency in Beijing-Tianjin-Hebei region. Especially since 2000, there is a remarkable trend. During the overall window period 24 years, the stability of energy efficiency in Hebei is better, the fluctuation of energy efficiency of Beijing and Tianjin is more obvious. Among them, Beijing's energy efficiency's rise is the highest, which is 2.77 percent. Tianjin comes second with 0.54 percent. The difference in energy efficiency in Beijing-Tianjin-Hebei region is shrinking, relying mainly on the spread and spillover of energy technologies.

5.2. Policy recommendations

Based on the above empirical analysis, this paper gives the following suggestions.

First, technological progress or innovation is the main source of energy efficiency improvement. There are specific ways to accelerate energy technology progress or innovation in

Beijing-Tianjin-Hebei region, such as do well in the introduction and training of energy technology talents, to expand the development and investment of energy technology facilities, and to strengthen the exchanges and cooperation between regional energy technologies.

Second, the upgrading and transformation of industrial structure is an important way to improve energy efficiency, including improving the proportion of the third industry in Beijing-Tianjin-Hebei region and increasing the use of new energy and clean energy.

Third, industrialization, urbanization and low-carbon lifestyle are effective means of improving energy efficiency. Such as improving the quality of infrastructure construction in Beijing-Tianjin-Hebei region, promoting green travel and energy conservation and emission reduction.

Beijing-Tianjin-Hebei region should formulate differentiated energy policies for local conditions, promote innovation in energy systems, adjust the proportion of energy consumption, deepen the reform of energy prices, optimize the allocation of energy resources and build a resource-conserving and environment-friendly society.

References

[1] Qu Xiaoe, Change of the total factor energy efficiency of Chinese province- An empirical study based on Malmquist index [J]. Research on quantitative economic and technical economics, 2009, (8) : 29-43.

[2] Wang Qunwei, Zhou Dequn, Shen Xuan, Measure and analysis of the total factor energy efficiency in China [J]. Journal of management review, 2010, 22 (3) : 37-43.

[3] Li Lanbing, China's total factor energy efficiency evaluation and deconstruction -- based on the dual perspective of "management - environment" [J]. China industrial economy, 2012, (6) : 57-69.

[4] Cheng Yuhong, Zhang Weiqi, The empirical analysis of the variation of energy efficiency in total factor energy efficiency in China's provincial regions -- taking guangdong province as an example [J]. Technical economy, 2013, 32 (3) : 56-63.

[5] Yang Lili, Shao Shuai, Cao Jianhua, Energy efficiency change decomposition and influencing factors of industrial whole factors in Yangtze river delta urban agglomeration -- empirical study on stochastic frontier production function [J]. Journal of Shanghai university of finance and economics, 2014, 16 (3) : 95-102.

[6] Chen Guanjv, Research on the energy efficiency and influencing factors of whole factors in China's manufacturing industry -- a stochastic frontier analysis based on panel data [J]. China soft science, 2014, (1) : 180-192.

[7] Charnes A.,W.W. Cooper,A.Y. Lewin,and L.M. Seiford. Extensions to DEA Models,Data Envelopment Analysis: Theory,Methodology,and Application[M]. Norwell,Massachusetts: Kluwer Academic Publishers,1994.

[8] Kevin Cullinane D -W S , Ping JI, Teng-Fei Wang. An Application of DEA Window Analysis to Container Port Production Efficiency[J]. Review of Network Economics, 2004, 3(2):184-206.

[9] Fare R,Grosskopf S,Norrism,etal. Productivity Growth,Technical Progress,and Efficiency Change in Industrialized Countries [J] . American Economic Review, 1994, (84):66-83.

Based on the Theory of TRIZ Solving the Problem of 18650 Battery Electrolyte Filling

CUI Shao-hua[1,*], MEI JIANG-ping[1], ZHANG Ling-hua[2] and DU Xiao[2]
[1]School of Mechanical engineering,Tianjin University,Tianjin,300072,
[2]Tianjin Lishen battery joint-stock company limited, Tianjin, 300384)

*csh_ls@126.com

Abstract. As a type of standardized battery cylindrical 18650 lithium-ion battery is widely used in new energy vehicle industry, It can be produced in large quantities without changing type. Because of its special advantages than others. But due to the pressure of rising capacity, electrolyte filling (which is short for E/L) process has become more and more difficult. While reducing the production efficiency eases the problem of E/L, it also poses performance and security problems. So the issue cannot be solved using the common knowledge of the industry. In this paper, This article does not use lean manufacturing or 6Sigma methods, we use TRIZ theory to analyze the E/L difficulty problem in detail (using causal analysis, technical contradiction analysis, substance – field analysis, physical contradiction analysis and other tools). By creating an atmosphere of vacuum and pressure replace the existing E/L tooling for single cell mechanical structure,through blowing hot air method to increase the temperature of electrolyte , Dissolving the J/R into a electrolyte tank which is full of 0.3Mpa nitrogen .Under the premise of not reducing the production efficiency, at the same time ensuring performance and safety, we try to find out a method to solve the E/L difficulty problem, and would get better application in the construction of new production lines in the new factory.

1. Introduction

TRIZ theory was established by G.S. Altshuller, a USSR inventor, in 1940s. The Russian counterpart of TRIZ is теории решения изобрет-ательских задач. Its phonetic transcription is Teoriya Resheniya Izobreatatelskikh Zadatch, which is the source of the short form of TRIZ [1]. It means the theory of solutions of invention issues. According to G. S. Altshuller, there are objective scientific principles and rules in the process of solving issues during invention and such principles and rules are repeatedly used in various inventions, only differing in technical fields of their application. Upon over 50 years of collecting, studying, sorting out, summarizing and extracting millions of patent literatures, a whole set of systemized and practical system of theories and method for solving invention-related problems is established.

Currently, TRIZ tools have been widely used in high-tech field. In this paper, TRIZ Theory would be used to solve practical issues in production of the E/L Procedure of 18650 Battery.

Cylindrical 18650 lithium-ion battery is a standard battery used in new energy vehicles industry, having the advantage of large-scale automated production, a single model, high safety factor, low cost etc. As the driving energy, under the premise of weight and space restrictions, it requires that the battery capacity is rising gradually. Because 18650 battery has a standard external dimensions (its structure is shown in Figure 1), the internal space is very limited, so that the battery need more and more positive and negative material and the amount of electrolyte liquid of positive and negative also increase, as a result the 18650 battery E/L process becomes more and more difficult. E/L difficulty problem will directly lead to the electrolyte's leaking, which would corrode the production equipment, and the electrolyte's leaking also result in lower amount of E/L that would cause the degradation of capacity and safety performance. To solve this problem, now it extends E/L time to settle, but the expense of productivity does not bring improvement in battery performance. Since the electrolyte prolong the time to expose in the air which would absorb moisture from the air, electrolyte composition will change, causing the battery self-discharge. Therefore it is posing an urgent need to address this problem.

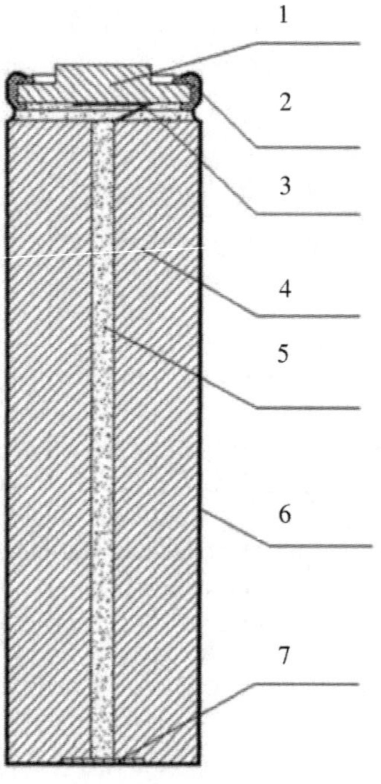

Figure 1. Battery 18650 structure

1:Cap;2:Seals;3:Positive tab;4:Positive/Negative electrode ;5:Free electrolyte;6: Battery shell;7:Negative tab

2. Analysis

As to 18650 E/L difficulty problems, we will use TRIZ theory to analyze and find the solutions.

2.1. Problems existence in current technology systems

Firstly, owing to the rising capacity, positive and negative material have been full of space inside the battery, it can only be evacuated the air within the cell as much as possible, and inject the electrolyte into the battery by pressurizing. Then repeatedly pressurizing and standing in order to inject the electrolyte into the battery. Secondly, the vacuumization mainly relies on E/L sealing fixture which may lose efficiency after many repeated times' opening, thus lower its effect of vacuumizing and pressurizing. Further, for E/L difficulty, it can be solved by repeatedly pressurizing and E/L filling and prolonging standing time, but all these will reduce productivity.

From the establishment of functional model (Figure 2), the power-driven pump make stored electrolyte in E/L tools into electrode plate, and the impact of the electrolyte to electrode plate is a short action. It is determined the relationship between the system and its components by functional model analysis, and found the functional factors of E/L difficult problems.

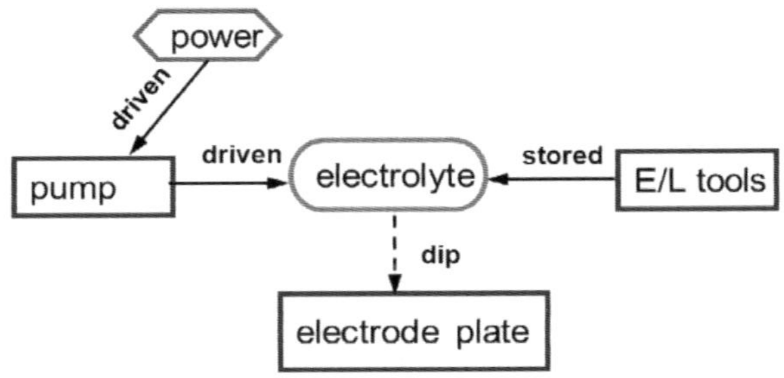

Figure 2. The establishment of functional model

2.2. Causal Analysis

As to E/L difficulty, we use causal analysis method (as shown in Figure 3). Through the analysis of the reasons step by step, we can draw a conclusion: the small space of positive and negative, the short time of E/L and the mass amount of E/L are the main reasons to E/L difficulty. These variables need to be further verified whether they are the key factors.

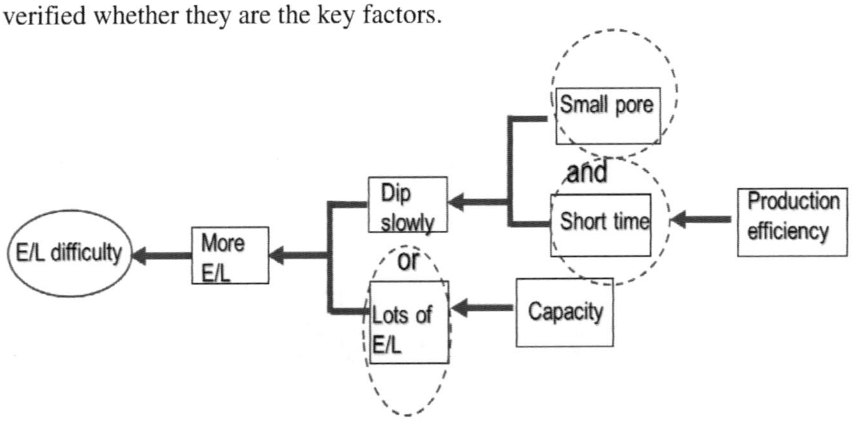

Figure 3. The causal analysis chart of 18650 battery E/L difficulty

Through causal analysis, we can determine the problem spots: 1: The pore between positive and negative material is too small for the electrolyte to enter. 2: The large amount of electrolyte has difficulty in entering into the jellyroll (which is short for J/R). 3: Due to the short time, electrolyte cannot be fully absorbed. Obviously, reducing the production efficiency is not an ideal solution, for which could enhance and eliminate both the beneficial and harm function. So the issue is not a regular problem that would not be solved within the expertise, we need to use TRIZ theory to solve it. Therefore we start our experiment with question 3.

2.3. Technical contradiction analysis

It can be determined that capacity and production efficiency are technical contradiction with the theory of technical contradiction analysis. While the E/L amount is increasing, so the capacity, the production efficiency is reducing; on the contrary, the E/L amount is reducing, so the capacity, the production efficiency is increasing. Therefore, we make E/L amount increase, capacity increase and production efficiency reduce as the main technology contradiction to solve [2] (as one can see in figure 4).

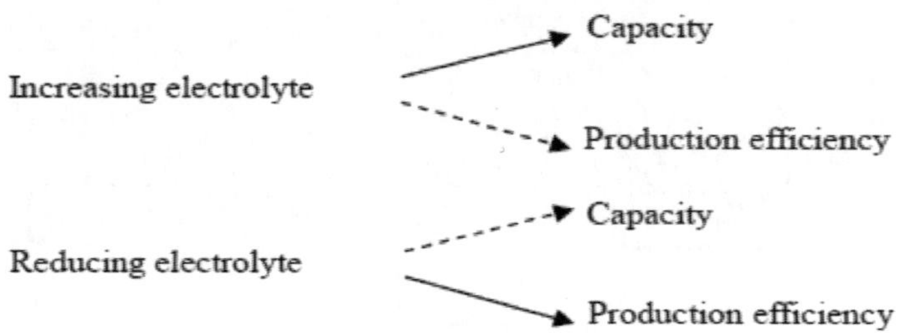

Figure 4. Technical contradiction model

Determine the engineering parameters: Improving parameters: Power; Deteriorating parameters: production efficiency. Based on the contradiction matrix, we obtained the following principles of the invention: No.28 the replacement of mechanical system, No.34 abandonment and repair, No.35 parameter changes. Then we come to the following plan: No.28 the replacement of mechanical system: creating an atmosphere of vacuum and pressure replace the existing E/L tooling; No.34 abandonment and repair: taping to form a cylinder on the mouth of the cell replace the existing E/L tooling, and after electrolyte absorption discard it; No.35 parameter changes: changing the electrolyte temperature to improve the absorption efficiency.

2.4. Substance field analysis

Through the substance field model, as shown in Figure 5, it can be seen that under the effect of dissolved chemical field force Fch, electrolyte S2 is a shortage that cannot enter into the J/R easily, so we introduced a new field Fch2 to improve the procedure E/L, then it would make the shortage to a standard effect.

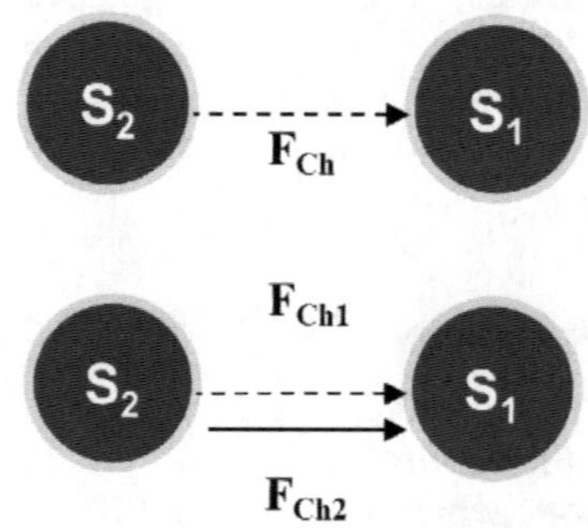

Figure 5. Substance field analysis model

S2: Electrolyte; S1: J/R; Fch: dissolved chemical field

According to the substance field model and the application of 76 standard solutions, the standard solution is obtained: No.2.25, No.3.1.2, No.3.1.4 [3]. Then we come to the following plan. Based on No.2.2.5 standard solution: by moving field instead of static field to give the solution of the problem as follows: high frequency alternation of vacuum and pressure to replace long time static vacuum and high pressure; Based on No.3.1.2 standard solution: changing the connections between the dual-system or multi-system to give the solution of the problem as follows: designing a new tooling, to ensure that no excess liquid floating instead of evacuated to vacuum and high-pressure, floating liquid is an obstacle; Based on No.3.1.4 standard solution: simplifying the solution of the dual-system and multi-system to give the solution of the problem as follows: putting large number of batteries into a vacuum and high-pressure tank to vacuumize and pressurize instead of the way used before.

2.5. Physical contradiction analysis

Through establishing a physical contradiction analysis model (Figure 6), it is found that the microstructure of the electrode plate must be both loose and compact. The electrolyte has easier access to the inner with loose microstructure, but too loose may cause cracking; so the electrode plate need to have dense microstructure, which would result in E/L difficulty. Therefore the electrode plate's microstructure is the physical contradiction [4].

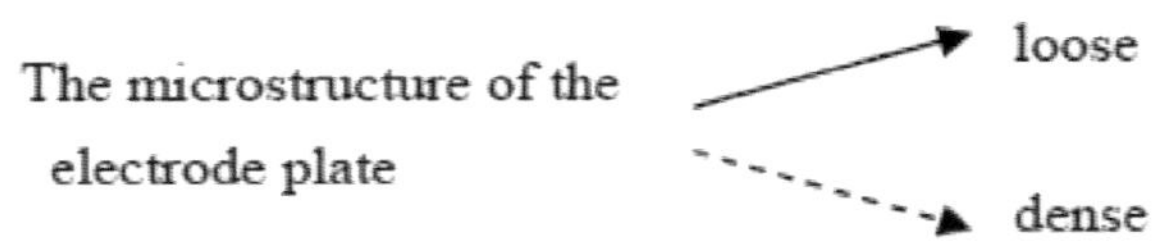

Figure 6. Physical contradiction analysis model

According to the physical contradiction analysis model and the application of 76 standard solutions, the standard solution is obtained: No.5.1.1, No.5.1.3. Then we come to the following plan. Based on No. 5.1.1 standard solution: the introduction of substance to give the solution of the problem as follows: when the electrolyte filled with nitrogen enters into the J/R, vacuumizing to expand the positive and negative pore making electrolyte easier to enter. Based on No. 5.1.3 standard solution: additives disappearing to give the solution of the problem as follows: mixing electrolyte composition in positive and negative material, after the electrolyte enter into the J/R, the composition melt, expanding the pore to make it easier to the J/R.

3. Results

This paper comes to three results.

1) To overall vacuumize and pressurize 1000 batteries by creating an atmosphere of vacuum and pressure replace the existing E/L tooling for single cell mechanical structure [5]. Thus ensuring the stability and reliability of the system, E/L time for single cell is longer than the existing time, but operating 1000 batteries at one-time, overall productivity is not reducing, but increasing.
2) Changing the electrolyte's temperature: in the experiment, through blowing hot air method to increase the temperature of electrolyte to 35 ~ 40 °C [6], the E/L absorption time is shortened by 14%, thereby the electrolyte absorption efficiency is improving.
3) Dissolving the J/R into a electrolyte tank which is full of 0.3Mpa nitrogen, the pore of J/R is expanded after vacuumization in order to create easier access for the electrolyte to get in, finally the J/R is back to normal as its pore shrink after pressurization.

4. Conclusions

Based on the TRIZ theory system, we have found a solution to solve the E/L difficulty of 18650 battery in production process. Product performance and security have improved, while the production efficiency is not reducing, and the system does not become complicated, but easier to operate and maintain. Improvement effect has experienced production test, and would get better application in the construction of new production lines in the new factory.

References

[1] Liu Xuntao. TRIZ Theory and Application[M], Beijing, Peking University publication, 2011:P53
[2] Zhao Min.TRIZ of approach and practice[M], Beijing,Cornell University Press,2009:P35
[3] Tan Runhua. Innovation Process and Methods of TRIZ and Application Technologies [M]. Beijing: Higher Education Press, 2012: P103
[4] Shen Menghong. Methods of TRIZ[M], Beijing,Peking University publication,2011:P137
[5] Cui Shaohua. Liquid filling device for battery [P].CN:2016208614561,2017:P3
[6] Cui Shaohua. Liquid filling device for round lithium ion battery [P]. CN:2016208614612, 2017:P5

Facile Preparation of Magnetic Graphene Oxide and Attapulgite Composite Adsorbent for the Adsorption of Ni (II)

WEI Bigui[1,*], ZHU Xiaofei[1] and CHENG Xiabing[1]

[1]School of Environmental and Municipal Engineering, Lanzhou Jiaotong University, Lanzhou, Gansu, PR China

*36935112@qq.com

Abstract. Graphene oxide (GO) is an excellent absorbent for heavy ion from wastewater, but it is hard to separate from water. To improve the adsorption capacity and separation performance of GO to nickel-containing wastewater, a composite magnetic GO-ATP adsorbent (MGA) was prepared by magnetizing GO and attapulgite (ATP) using ferroferric oxide and then carrying out hydrothermal reaction. The adsorption capacity and mechanism of MGA were investigated based on Ni^{2+} as targeted pollutant. Experimental results showed that the pH value significantly affects the removal rate of Ni^{2+}, which is mainly due to that OH^- in wastewater reacts with Ni^{2+}, resulting in sediment that leads to the increase of removal rate. MGA can achieve max adsorption capacity of Ni^{2+} to 190.8 mg/g at pH = 5, and the adsorption process was mainly determined by chemical adsorption, which was in line with pseudo-secondary dynamics model. The adsorption was basically homogeneous monolayer adsorption with heat release, which was more agree with Langmuir adsorption isotherm equation. the adsorption process of Ni^{2+} by MGA. The adsorption process was a spontaneous process and an exothermic reaction. It can be confirmed that the prepared MGA adsorbent can realize slurry separation using magnetic separation principle and has high adsorption capacity to Ni^{2+}.

1. Introduction

Graphene oxide GO contains several oxygen-containing functional groups including hydroxy group (-OH), epoxy group (-C=O), carboxy group (-COOH), with high specific surface area [1]. Compared with other carbonaceous materials, GO has better environmental friendliness and biological compatibility [2]. Oxygen-containing functional groups can serve as adsorption sites for heavy metal ions in wastewater [3]. Due to the binding action between oxygen-containing functional groups and heavy metal ions, the heavy metal ions in wastewater can be removed[4], so that the purpose of wastewater purification can be realized. In addition, due to the existence of these oxygen-containing functional groups, GO can be steadily distributed in water or other polarity organic solvents[5]. However, according to current literature reported on pure GO or GO-based absorbent, harsh preparation conditions, complex preparation method, and hard separation and high cost are often involved. Therefore, inorganic clay minerals with excellent adsorption and separation capacities were adopted to be coupled with GO, so as to prepare heavy metal ions absorbent with excellent property and lower price.

At present, there are many domestic and foreign researches on GO composite material, while there has been no researches on composite of GO and inorganic clay attapulgite (ATP). GO is in two-dimensional plane form, while a single crystal of ATP is in fibriform [6].

This study aims to prepare a novel composite in "plane-needle" structure by inserting needle-like ATP on GO surface, so as to improve GO property and reduce the cost in treatment of wastewater containing heavy metals.

To improve separation performance, nano-Fe_3O_4 was used to magnetize absorbent, so that the slurry can be separated by magnetic attraction. In addition[7, 8], the adsorption property and mechanism were investigated using common heavy metal pollutant Ni^{2+} as target pollutant.

2. Materials and methods

2.1. Materials

The 2mg/mL aqueous solution of graphene oxide was purchased from Nano technology Co., *Ltd.* ATP was purchased from Anhui Guangming Attapulgite Clay Plant. Other reagents were of analytical purity or excellent purity. Deionized water was prepared by in laboratory. Ni^{2+} stock solution was prepared by nickel chloride configuration. The pH was adjusted by 0.1mol/L HCl solution or 0.1mol/L NaOH solution.

2.2. Methods

2.2.1 Preparation of composite absorbent of magnetic graphene oxide and attapulgite. Pour 2mg/L GO aqueous solution into high temperature reaction kettle, and then add certain amount of ferroferric oxide, let them react at 200℃ for 8h. Then, remove supernatant liquor, dry and grind residue, resulting in magnetic GO powder. Add certain amount of magnetic GO into 10% acetic acid solution while keeping magnetic stirring for 10min, resulting in dispersion system of magnetic GO. Similarly, add certain amount of ATP into methanol solution while keeping magnetic stirring for 10min, resulting in ATP dispersion system. Subsequently, mix the two systems by certain ratio, keep magnetic stirring for 24 at room temperature, stand for 3h, remove supernatant liquor, then dry the residue at 150℃ for 5h to obtain solid product, and grind it into powders, which is the final product-composite absorbent of magnetic GO-ATP (MGA).

2.2.2 Adsorption experiment. Add certain amount of MGA into Ni^{2+} aqueous solution of different initial concentrations and different pH values, and then transfer the mixture solution into a conical flask for oscillatory reaction at 250r/min for certain period of time. After that, place the conical flask on a magnet, let it stand for 3min, remove supernatant. Then, atomic absorption spectrophotometer (Spectr AA110/220, Varian, United States) was used to measure the concentration of Ni^{2+}. The average value of 3 parallel experiment results was adopted as the final result. The removal rate of Ni^{2+} can be obtained by the equation below:

$$\eta = \frac{C_0 - C_e}{C_0} \times 100\% \tag{1}$$

where C_0 is Ni^{2+} initial concentration, and C_e is Ni^{2+} equilibrium concentration, respectively, mg/L; η represents removal rate, %.

3. Result and discussion

3.1. Effect of pH

The C_0 of Ni^{2+} was 100mg/L, the concentration of MGA absorbent was 1.4g/L, and the removal rates of Ni^{2+} at different pH values are shown in Figure 1 From Figure 1 we can know that the removal rate of Ni^{2+} by MGA increases with pH value. Within the pH value range of 2-7, the removal rate increases slowly, which indicates that pH value under acid condition does not significantly affect removal rate; Within the pH value range of 7-10, the pH value significantly affects the removal rate of Ni^{2+}, which is mainly due to that OH^- in wastewater reacts with Ni^{2+}, resulting in sediment that leads to the increase of removal rate. To remove the influence of OH^- to experiment results, pH value was determined to 5 in follow-up studies.

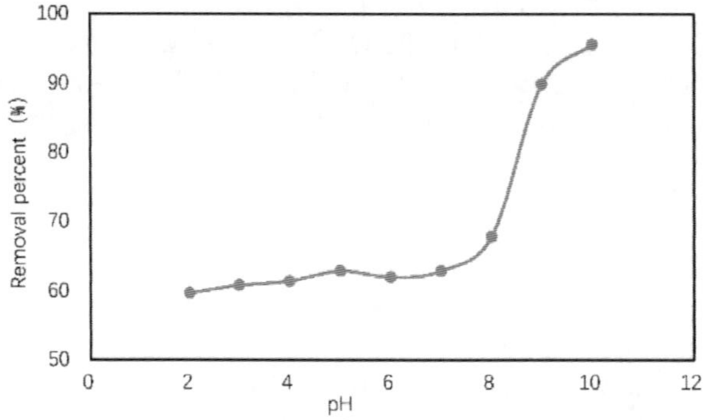

Figure 1. Effect of pH of Ni^{2+} adsorbed on MGA.

3.2. Adsorption kinetics

Adsorption mechanism can be analyzed based on two kinetic models. The pseudo-first order kinetic model is shown as below[9]:

$$\log(q_e - q_t) = \log q_e - \frac{k_1}{2.303}t \tag{2}$$

$$q_t = \frac{(C_0 - C_e)V}{m} \tag{3}$$

In both above equations, q_t is the adsorption capacity at time *t*, mg/g; q_e is equilibrium adsorption capacity, mg/g; k_1 is rate constant, min^{-1}; *m* is the mass of absorbent, g; *V* is the volume of adsorption system solution, L.

The pseudo-second order kinetic model is shown as below [10]:

$$\frac{t}{q_t} = \frac{1}{k_2 q_e^2} + \frac{1}{q_e}t \tag{4}$$

where k_2 is the rate constant of second-order kinetics, g/(mg·min).

To carry out nonlinear fitting, equation (3) and (4) was rewritten into non-linear form:

$$q_t = q_e(1 - e^{-k_1 t}) \tag{5}$$

$$q_t = \frac{k_2 q_e^2 t}{1 + k_2 q_e t} \tag{6}$$

The nonlinear fitting results of pseudo-first order kinetic model and pseudo-second order kinetic model [11] are shown in Figure 2. The fitting parameters are summarized in Table 1.

Table 1. Calculated kinetic parameters for Ni^{2+} adsorption using MGA.

Pseudo-first order model			Pseudo-second order model		
K_1 (min^{-1})	q_e(mg/g)	R^2	K_2 (g/(mg·min))	q_e(mg/g)	R^2
0.0692	46.58	0.9421	0.0020	50.68	0.9962

According to Table 1, it can be known that the adsorption of Ni^{2+} by MGA is more in line with pseudo-second order kinetics, with R^2=0.996, which indicates that the adsorption rate is mainly determined by chemical adsorption process [12]. This result reveals that such adsorption reaction involved electron transfer and exchange and the formation of chemical bonds. Within the first 5 min, the rate of chemical adsorption reaction is quick, leading to a significant acceleration of adsorption rate.

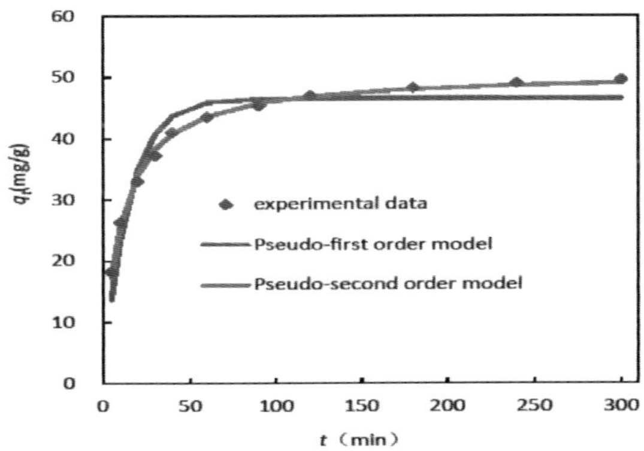

Figure 2. Adsorption kinetics of Ni^{2+} on MGA.

3.3. Adsorption isotherm

The Langmuir and Freundlich isotherm adsorption model was used to fit the experimental data.
As a classic adsorption model proposed by I. Langmuir in 1918 [13], Langmuir model assumes that the absorbent surface is fully homogenous, i.e. all adsorption sites have identical adsorption energy, and only monolayer adsorption took place for adsorbate. Langmuir model has been extensively used to describe various adsorption phenomena in gas and solution, of which the equation is expressed as follow:

$$q_e = \frac{q_{max} K_L C_e}{1 + K_L C_e} \tag{7}$$

where K_L is equilibrium adsorption constant, which reveals the relation between Ni^{2+} binding site and affinity; q_{max} is the max adsorption capacity of Ni^{2+} by composite absorbent, mg/g.

Being proposed by H. M. Freundlich in 1906 [14], Freundlich model is another classic frequently-used adsorption model. Freundlich model assumes that the absorbent surface is inhomogeneous, i.e. the energies of adsorption sites are not identical but distributed in exponential curve. Freundlich model equation is shown as below:

$$q_e = K_f C_e^{\frac{1}{n}} \tag{8}$$

where K_f represents Freundlich empirical constant, n is surface heterogeneity coefficient.

The nonlinear fitting results of adsorption isotherms for different initial concentrations of Ni^{2+} at 298K are shown in Figure 3. The fitting parameters are summarized in Table 2.

Table 2. Parameters of two adsorption isotherm models for Ni^{2+} on MGA.

Langmuir			Freundlich		
q_{max}(mg/g)	K_L(L/mg)	R^2	K_F (mg/g)	$1/n$	R^2
190.8	0.046	0.9932	18.38	0.500	0.9719

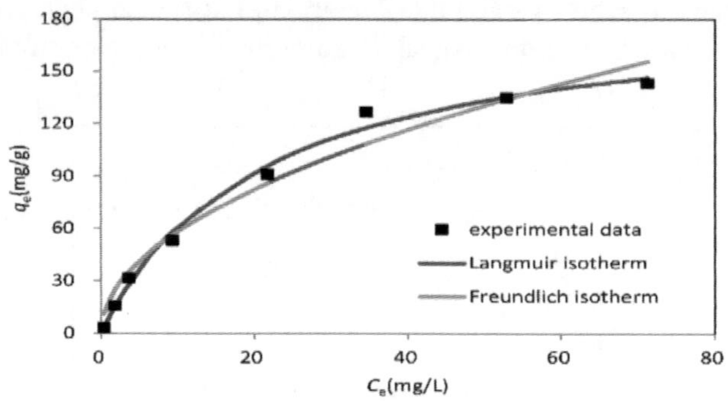

Figure 3. Adsorption isotherm models for Ni^{2+} on MGA.

3.4. Adsorption thermodynamics

The adsorption thermodynamics parameters (ΔH^0, ΔG^0, and ΔS^0) of Ni^{2+} on absorbent surface can be calculated by temperature-dependent thermodynamic equilibrium constant K_0. The calculation equation of standard free energyΔG0 is shown as below [15]:

$$\Delta G^0 = -RT\ln K_0 = \Delta H^0 - T\Delta S^0 \tag{9}$$

where T is absolute temperature, K; K_0 is the thermodynamic equilibrium constant in mg/L, which can be calculated by $\frac{q_e}{C_e}$; R is universal gas constant, $8.314\text{J}/(\text{mol}\cdot\text{K})$. According to equation (9), the standard entropy change (ΔS^0) and standard enthalpy change (ΔH^0) can be calculated according to the liner equation below:

$$\ln K_0 = \frac{\Delta S^0}{R} - \frac{\Delta H^0}{RT} \tag{10}$$

The adsorption thermodynamics parameters of Ni^{2+} on absorbent surface are shown in Table 3.

Table 3. The parameters of thermodynamic for the Ni^{2+} adsorption on MGA.

T (K)	ΔG^0(kJ/*mol*)	ΔS^0(J/(mol·K))	ΔH^0(*kJ*/mol)
298	-5.37		
308	-5.88	-8.71	0.05
318	-6.13		

The thermodynamic parameters of adsorption reaction is not only associated with adsorbate property, but also with absorbent property. According to Table 3, the ΔG^0 less than 0 and ΔH^0 greater than 0, which indicates the adsorption process of Ni^{2+} by MGA was a spontaneous process [16, 17]. The ΔS^0 is negative illustrates that adsorption process of Ni^{2+} by MGA was am exothermic reaction [18].

4. Conclusion

MGA is a simple and low-cost heavy metal ions absorbent, which can be prepared by simple magnetization and hydrothermal reactions. A good slurry separation effect can be realized by magnetic force. The prepared MGA absorbent had large adsorption capacity of Ni^{2+}, and it was a spontaneous adsorption process. The prepared MGA absorbent can be used in treatment of heavy metal wastewater, which enjoys certain potential of application.

Acknowledgments

This work was supported by the Foundation of A Hundred Youth Talents Training Program of Lanzhou Jiaotong University (No. 152022) and Natural Science Foundation of Gansu Province of China (No. 145RJZA209).

IOP Conf. Series: Earth and Environmental Science **104** (2017) 012019 doi:10.1088/1755-1315/104/1/012019

References

[1] D.R. Dreyer, S. Park, C.W. Bielawski, R.S. Ruoff, The chemistry of graphene oxide, Chemical Society Reviews, 39 (2010) 228-240.
[2] L. Chen, P. Hu, L. Zhang, S. Huang, L. Luo, C. Huang, Toxicity of graphene oxide and multi-walled carbon nanotubes against human cells and zebrafish, Science China Chemistry, 55 (2012) 2209-2216.
[3] Z. Wu, H. Zhong, X. Yuan, H. Wang, L. Wang, X. Chen, G. Zeng, Y. Wu, Adsorptive removal of methylene blue by rhamnolipid-functionalized graphene oxide from wastewater, Water research, 67 (2014) 330-344.
[4] X. Zhu, Y. Cui, X. Chang, H. Wang, Selective solid-phase extraction and analysis of trace-level Cr (III), Fe (III), Pb (II), and Mn (II) Ions in wastewater using diethylenetriamine-functionalized carbon nanotubes dispersed in graphene oxide colloids, Talanta, 146 (2016) 358-363.
[5] S. Luo, X. Xu, G. Zhou, C. Liu, Y. Tang, Y. Liu, Amino siloxane oligomer-linked graphene oxide as an efficient adsorbent for removal of Pb (II) from wastewater, Journal of hazardous materials, 274 (2014) 145-155.
[6] A. Li, A. Wang, J. Chen, Studies on poly (acrylic acid)/attapulgite superabsorbent composite. I. Synthesis and characterization, Journal of Applied Polymer Science, 92 (2004) 1596-1603.
[7] J.-h. Ke, Z.-k. Wang, Y.-z. Li, Q.-l. Hu, J. Feng, Ferroferric oxide/chitosan scaffolds with three-dimensional oriented structure, Chinese Journal of Polymer Science, 30 (2012) 436-442.
[8] A. Mostafa, A. El-Dissouky, A. Fawzy, A. Farghaly, P. Peu, P. Dabert, S. Le Roux, A. Tawfik, Magnetite/graphene oxide nano-composite for enhancement of hydrogen production from gelatinaceous wastewater, Bioresource technology, 216 (2016) 520-528.
[9] D.J. O'Shannessy, D.J. Winzor, Interpretation of Deviations from Pseudo-First-Order Kinetic Behavior in the Characterization of Ligand Binding by Biosensor Technology, Analytical Biochemistry, 236 (1996) 275-283.
[10] Y.-S. Ho, Second-order kinetic model for the sorption of cadmium onto tree fern: A comparison of linear and non-linear methods, Water Research, 40 (2006) 119-125.
[11] F.M. de Oliveira, B.F. Somera, E.S. Ribeiro, M.G. Segatelli, M.J. Santos Yabe, E. Galunin, C.s.R.T. Tarley, Kinetic and isotherm studies of Ni2+ adsorption on poly (methacrylic acid) synthesized through a hierarchical double-imprinting method using a Ni2+ ion and cationic surfactant as templates, Industrial & Engineering Chemistry Research, 52 (2013) 8550-8557.
[12] L. Cui, Y. Wang, L. Gao, L. Hu, L. Yan, Q. Wei, B. Du, EDTA functionalized magnetic graphene oxide for removal of Pb(II), Hg(II) and Cu(II) in water treatment: Adsorption mechanism and separation property, Chemical Engineering Journal, 281 (2015) 1-10.
[13] I. Lanowix, The adsorption of gases on plane surface of glass, mica and platinum, J. Am. chem. soc, 30 (1918) 1361.
[14] U. Freundlich, Die adsorption in lusungen, (1906).
[15] L. Liu, C. Li, C. Bao, Q. Jia, P. Xiao, X. Liu, Q. Zhang, Preparation and characterization of chitosan/graphene oxide composites for the adsorption of Au(III) and Pd(II), Talanta, 93 (2012) 350-357.
[16] R. Niwas, U. Gupta, A.A. Khan, K.G. Varshney, The adsorption of phosphamidon on the surface of styrene supported zirconium (IV) tungstophosphate: a thermodynamic study, Colloids and Surfaces A: Physicochemical and Engineering Aspects, 164 (2000) 115-119.
[17] Y. Wang, X. Wang, X. Wang, M. Liu, L. Yang, Z. Wu, S. Xia, J. Zhao, Adsorption of Pb (II) in aqueous solutions by bamboo charcoal modified with KMnO 4 via microwave irradiation, Colloids and Surfaces A: Physicochemical and Engineering Aspects, 414 (2012) 1-8.
[18] Y.-H. Li, Z. Di, J. Ding, D. Wu, Z. Luan, Y. Zhu, Adsorption thermodynamic, kinetic and desorption studies of Pb2+ on carbon nanotubes, Water Research, 39 (2005) 605-609.

Rural Design Ethics Based on Four Dimensions

Jiaxin Zhang[1,*] and Li Zhu[1]

[1]Room410, No.405 Xiao Xiang Road, Yu Lu District, Changsha China

*jiaxinzhang@csu.edu.cn

Abstract. China has a large rural area with a large population, whose architectural features, natural landscape, organizational structure and industrial structure are very different from that of cities. In the past, the contradictory between city and rural areas in China had negative effects on rural construction, resulting in a slow development. The excessive focus on city design has led to the neglect of rural design. Blindly using the concept and method of city design to renewed the countryside is a kind of destruction to the countryside, and also wastes a lot of construction resources. Design is influenced by ethical concepts, which needs to pay more attention to the culture tendency and society. Urban design makes theoretical investigation aiming at the ethical questions that emerged from city, then summarizes the design strategies of the city. While Chinese rural design has only begun to enter people's horizon, and there is very little discussion about it. Due to the lack of ethical value guidance, Chinese rural design and construction has many problems at different levels of ecology, culture and industry. Therefore this paper primarily explores the domestic and foreign design ethics, attempting to provide a new perspective for Chinese rural design, aiming at finding a realistic and forward-looking solution for Chinese rural design concerning to the complex relation between city and rural areas.

1. Introduction

The process of urbanization in China intensifies the decline and disappearance of traditional villages. Therefore, it is urgent to strengthen the protection and development of traditional villages. In view of the prominent contradictions and drawbacks in the rural construction, it is necessary to guide the village construction with the design ethics. This paper hopes to reflect on the historical reality from the perspective of design ethics, learn from the current situation of domestic and foreign country design research.However, due to the domestic and aboard research on the traditional village design ethics has just started, there are many problems to be solved in theory and practice.

Most foreign rural villages have experienced the development process of city and rural conflict,so that a considerable amount of experience had been summarized in theory and design methods, which provide reference for China's rural design.

Domestic scholars are very concerned about rural issues, but they may lack a comprehensive, in-depth and objective judgment, whether they are in rational exploration or practical exploration. The contradiction of urban and rural binary opposition has yet to be alleviated, so in the face of the issues, such as rural cultural destruction, the ecological environment destruction and lagging economy, we must establish a kind of design concept based on ethical care, and then from the ecology, the industry, the people's livelihood and the spirit dimension, we have to make the strategic thinking for rural design at present and in the future.

2. Abroad Research Status

In foreign country, design ethics has always been a very important issue. From the strong democratic thought of Bauhaus to Victor Papanek's design philosophy morality in Design for the Real World (1985) and then to the current concept of ecological design, as well as the concept of sustainable development can be summarized as some certain ideas repetition. Although, these theories have been discussed in depth, but there is no more profound expansion. All of these studies mainly aim efforts at criticizing city design, which talk less about traditional village design and ethics in the correspondence and the feasibility exploration of development of traditional rural design ethics.

At present, the study of the rural design ethics can be divided into four aspects:

2.1. The justice of rural environment

There are series of books refer to design justice such as *Justice, Nature and the Geography of Difference* (Harvey 2010) [1], *Design for the Real World* (Papanek 2016) [2], *Environment Ethics. Duties to the Value in the Nature World* (Rolston 2012) [3]. They comprise the environmental space, society, political and culture, and emphasis the living crisis of peasants caused by 'design injustice'. Papanek especially mentioned that design should be served for weak group. As a statement of equality, his consideration focused on social disadvantages of traditional villages.

Arendt and Brabec (1994) [4] mentioned that one-sided emphasis on industrialization was not the best choice in efficiency or fairness, and the scientific and comprehensive plan of rural area was the best choice for the countryside to get overall development. These researches mainly discussed the role of design in the social context from the perspective of ethics. They distinguished the rural design from city design, gave rural construction more impartiality and humanity thinking, and reflected deeper consideration of practice subject on traditional village.

2.2. The ecology of rural design

Architect PaoloSoleri combined Ecology with Architecture to establish a conceptual framework for "Ecological Architecture". (Olgyay et al., n.d.) [5] he put forward that architecture design should coordinate with regional geography and climate. (McHarg 1994) [6], raised an idea that people, architecture, nature and society should be coordinately developed. He explores the effective ways and methods of ecological architecture in rural design, which pays more attention to save energy and protect natural environment. The above theory discussed the sustainable design development from the design ethics' perspective. In addition, the practice of rural construction abroad focuses on the combination, neither design nor natural itself.

2.3. Rural design responsibility and moral transmission

Whiteley (2012)[7] emphasized designer's responsibility. Jonas (2000) [8] researched on the social responsibility of designers from the perspective of professional ethics. In 1987, American designer Ronald L.Mace began using the concept 'Universal Design' in large numbers, his point of view highlighted the moral transmission of design. In addition, rural construction practice has succeeded in foreign countries such as the ordinances of rural design promulgated in England, the rural community planning in Germany and the eco-village organization formed in Denmark, which emphasized the importance of design responsibility in the construction and protection of villages. The rural design practice in foreign countries concerned positive change in village, and how to convey moral education in design.

2.4. The economic value of rural design

Arendt R, in his works, talked extensively about the important influence of rural design on regional economic development, and rural design must include present situation of land resource using in this areas, as well as economic requirements of future development. (Gallent, n.d.) [9] they introduced the economic, community and landscape challenges faced by rural planning. He investigated the relationship between rural planning and rural economic affairs, and considered the inseparability of economic changes and changes of rural society and environment.

Foreign rural design explored the ways how to meet the economic requirements through designing village space, and how to deal with the conjunction of rural construction, sustainable development and regional cultural protection.

3. Domestic Research Status

In the drawbacks view of the traditional village construction under the impact of urbanization in China, the domestic scholars also discussed practical necessity of 'design ethics'.

Studies on the design ethics of domestic village constructions and protection mainly signalize in the following points:

- The design of modern technology localization. Namely, the design not only satisfies its modern functionality, but also maintains the traditional geographical and cultural characteristics.
- Respecting ecological environment includes the ecology of nature, culture and humanity.
-
- The reconstruction of value system: design injects norms and ethics into objects. Therefore, the "ceremony" and "morality" reflected from design penetrating into rural life will inject new vitality into rural areas. In essence, the "ceremony" designs the ethical entity and stipulates the relationship between people, reflects the ethical spirit of the individual returning to the entity.

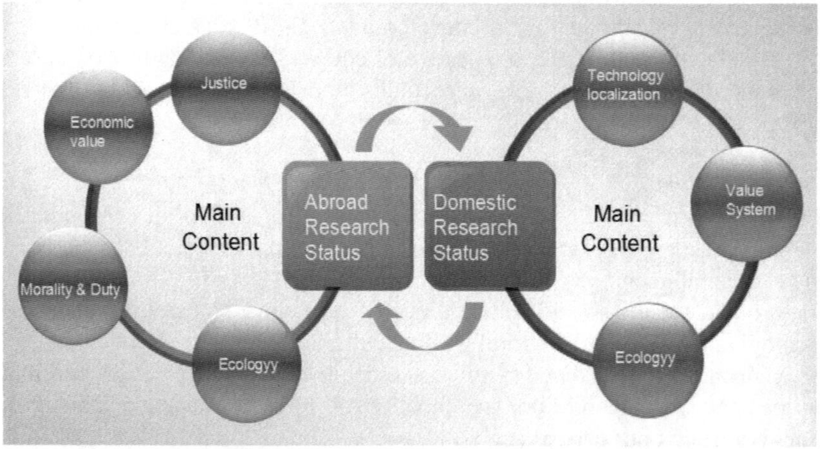

Figure 1. Domestic and international research reviews structure chart

The relevant researches at domestic and foreign research confirm the possibility of introducing 'design ethics' which emphasizes spatial justice, ecology, social responsibility and economic value into rural construction layout, however, there is a need for further research. First of all, there is no standardized design theory to guide the current rural constructions; secondly, the fundamental task of rural constructions should be cultural construction and the function of design ethics to the village culture cannot be ignored. Finally, the idealization of the relationship between design ethics and designers is confused. Therefore, further analysis should be carried out on economic and social context of ethical practice. Thus, a research on rural design ethics is in urgent need. As show in figure.1.

4. Four dimensions of rural design ethics

Rural areas are difficult to be a sustainable and updated entirety because of the lack of systematic theory and practical guidance, which brings about lots of problems in rural constructions. Aiming at the problems of homogenization and roughening in rural designs, the thesis systematically analyses the theoretical guidance to the overall design in rural areas through four dimensions.

4.1. Ecological dimension:

In the process of urbanization, the extensive economic development model leads to the shortage of ecological ethics in rural areas. The developmental pattern of 'getting rich first and then dealing with the pollution' obtains economic benefits at the cost of environment, which results in the waste of material resources in rural areas and lots of rubbish that is difficult to be decomposed in a short time. Furthermore, the obsolete production and consumption concept, as well as the insufficiency of ecological consciousness of famers lead to the destruction of ecological environment in rural areas. For example, many farmers bury or burn the discomposed domestic garbage without any environmental treatment and use excessive chemical fertilizers and pesticides. To solve the problems above, firstly, we can improve villagers' ecological and environmental protection awareness by learning their daily life and farm work habits. Then, in order to maintain the regional architecture characteristics as far as possible on the basis of respecting the original local ecology, rural environment can be changed by the means of ecological design and appropriate technology, as well as any other related approaches. Lastly, we must promote the ecological values of sustainable development and never drain the pond to get all the fish in the process of economic development. Meanwhile, the emphasizing on the balance mechanism of rural development and the rural ecological ethics projection to society, economy and culture can provide value guidance of ecological ethics in the constructions of beautiful countryside.

4.2. Industry dimension:
To coordinate development between city and rural areas, we should change the dual econo-mic structure in the countryside and the cities. The process of urbanization in China has led to a large number of rural human resources and material resources output to big cities, and the imbalance of rural industrial design has exacerbated the problem of rural hollowing. In rural constructions, industrial development is the fundamental driving force to promote the modernization process of rural areas, and also is the material basis of rural construction and development. At present, the rural complex formats have influence on the diversification of rural production mode and many aspects like rural population base, living standards, housing construction and infrastructures etc. Industrial dimension, based on respecting for cultural attribute and regional characteristics, upgrades the rural industry chain through the packaging design of agricultural products, handicraft products design, and consumption space design to create the industry platform with rural characteristics and boost the economic development.

4.3. People's livelihood dimension
The dimension of people's livelihood is the dynamic representation of village design ethics. It pays attention to public service security facilities like rural education, medical, transportation, energy, cleaning and the governance mechanisms in line with its own characteristics such as rural community and rural autonomy. The dimension of people's livelihood attaches great attention to the cultural protection of traditional villages, and provides the ethics guidance to the modern life style, appeared in rural areas, as well as critically inherits the traditional values and customs, which represents the concentration in the rural construction of the living environment and life quality.

4.4. Spiritual dimension:
Spiritual dimension is the core connotation of village design ethics. The stability and closeness of Chinese rural areas enable the traditional context with regional attributes to be preserved, and both have certain ethical enlightenment content. For example, the ethical and spiritual belief space like the village's clan temple, genealogy, inscriptions, nave, ancestral temple, simple shrine continually educate people through worship gods and their ancestors. Public order and good customs of the village help to constraint villagers' 4ehaviour. The villagers' self-confidence in village culture helps them consciously undertake and inherit the parental guidance and proverbs and regard them as rural value standard. The backwardness of rural development has led to the backwardness of farmers' values, and the rigid rural life causes the decline of farmers' personality, likewise all kinds of ideas updates slowly lead to the lack of an active humanistic impetus.

From the spiritual dimension, rural design extensively studies the history, culture and customs of countryside, and penetrates regional cultural spirit into the rural landscape architecture and daily life by the means of design, which takes the historical responsibility to continue of rural cultural value.

5. Summary

The rural design ethics abandoned the design mentality of simple visual sense and sensory stimulation to pursue the comprehensive design in ecology, industry, people's livelihood and spirit, so that the ethical significance of rural design has a broader meaning. As show in figure 2.

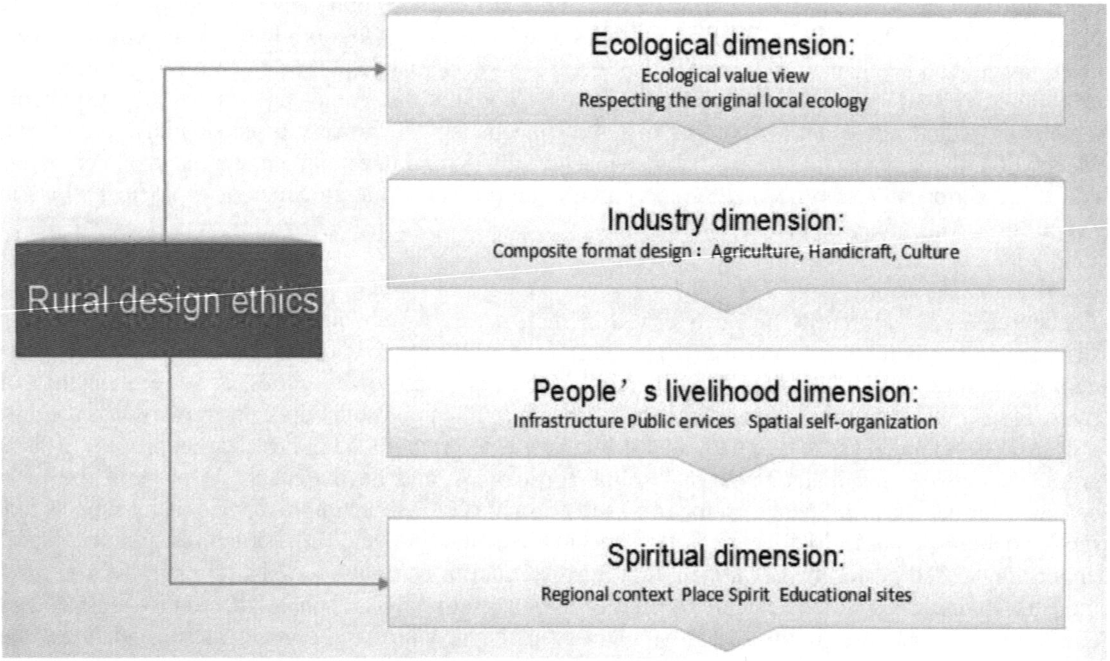

Figure 2. The key words of rural design ethics in four dimension

In the process of rural construction, designers must consider people how to live in ethical way and obtain a space for rural further development in all aspects by the means of design. In this way, village development will turn into an updating and sustainable organic integrity through the four dimensions.

References

[1] Harvey, D. (2010). Justice, nature and the geography of difference. Cambridge: Blackwell Publish.
[2] Papanek, V. (2016). Design for the Real World. London: Thames & Hudson.
[3] Arendt, R. and Brabec, E. (1994). Rural by design. Chicago, Ill.: Planners Press, American Planning Association.
[4] Rolston, H. (2012). Environmental Ethics. Philadelphia: Temple University Press.
[5] Olgyay, V., Olgyay, A., Lyndon, D., Olgyay, V., Reynolds, J. and Yeang, K. (n.d.). Design with climate.
[6] McHarg, I. (1994). Design with nature. New York: J. Wiley.
[7] Whiteley, N. (2012). Design for society. Johanneshov: TPB.
[8] Jonas, H. (2000). The imperative of responsibility. Chicago (Ill.): The University of Chicago Press.
[9] Gallent, N. (n.d.). Introduction to rural planning.

Institute of Physics
Dirac House, Temple Back
Bristol BS1 6BE UK

ISSN: 1755-1307
ISBN 978-1-5108-5493-2
9 781510 854932

CURRAN ASSOCIATES INC.
proceedings
.com

6th International Conference on Electronic Devices, Systems, and Applications (ICEDSA 2017)

IOP Conference Series: Materials Science and Engineering
Volume 340

Kuching, Malaysia
7 – 8 August 2017

Institute of Physics